Advance Praise

This important new study of the Civil War demonstrates that while blockade running was indeed the "lifeline of the Confederacy," the Union blockade and the capture of Confederate ports choked that lifeline to a small fragment of the South's needed seaborne commerce and played a key role in eventual Northern victory. Of special value is Gil Hahn's analysis of the naval war in the context of international law.

—James M. McPherson, author of *War on the Waters: The Union and Confederate Navies, 1861-1865*

More than a century and a half after its close, the Civil War's naval story is still little known or appreciated, particularly those clandestine adventures of the blockade runners that dared everything to smuggle foreign goods and munitions into the Confederacy, and the constant efforts of Union fleets trying to close that trade and starve the South. In *Campaign for the Confederate Coast*, Gil Hahn presents a strong argument for the centrality of this blockade war in determining the fate of the Confederacy, and the nation.

—William C. Davis, author of *The Greatest Fury: The Battle of New Orleans and the Rebirth of America*

In his *Campaign for the Confederate Coast*, Gil Hahn tackles all the complicated legal, logistical, and strategic factors entailed in the blockade imposed by the U.S. Navy on the Confederate States of America. Hahn ably covers such varied topics as the evolution of steam technology and ironclads, tactical developments in blockade-running, questions of supply, the effectiveness of Confederate efforts to counter a growing number of Federal ships, and the intricacies of international law.

Without getting bogged down in unnecessary detail, he skillfully summarizes the pertinent battles and campaigns, forgotten ships and international incidents, and the actions of naval and army officers and their civilian superiors. In his final pages, he reviews statistics and individual testimony to assess the blockade's role in the ultimate U.S. victory. This succinct yet comprehensive volume deserves a place on the bookshelf of every Civil War enthusiast.

—William W. Bergen, Independent Civil War scholar based in Charlottesville, Virginia

Gil Hahn's new survey of the Civil War on the Atlantic and Gulf coasts is more than just the conventional account of the U.S. Navy's blockade of the Confederacy. It embraces ship and weapons technology, coastal fortifications, charts and data on coastal sailing vessels, even sailors' rations. This is the single best survey of the conflict for the Confederacy's coastlines on offer, and its conclusions will surprise both the skeptics and the advocates of the blockade's effectiveness.

—Allen C. Guelzo, Princeton University, author of *Gettysburg: The Last Invasion*

Gil Hahn's book gives a clear and concise account of the Federal Navy's operations during the American Civil War and of the Confederate government's attempts to conduct trade and counter the blockade. His work is among the best I've read at explaining the details of nineteenth century naval campaigns, particularly what it took to keep ships and their crews combat-ready during complex and far-reaching missions. I highly recommend Gil Hahn's book to anyone looking to understand the big picture of American Civil War naval operations.

—Lucas Clawson, Hagley Historian, Hagley Museum and Library, Wilmington, Delaware

Breaking the mystique of the blockade has eluded scholars until the publication of *Campaign for the Confederate Coast*. Based on meticulous research, written in a lively prose, and judiciously argued, Gil Hahn has produced an immensely important book that illustrates the ways that the Federal Navy put the squeeze on the South by choking off the flow of essential goods for the waging of war. Hahn also does an impressive job of capturing how Confederate blockade runners were able to elude their Northern pursuers in dramatic operations that helped, in no small way, to prolonging the Civil War.

—Peter S. Carmichael, Director of the Civil War Institute at Gettysburg College

Campaign for the Confederate Coast: Blockading, Blockade Running and Related Endeavors During the American Civil War clearly and cogently describes, from both Southern and Northern points of view, the dozens upon dozens economic, technology and military policy conditions and adaptations which created military outcomes of the war. Thoroughly, Hahn offers this in a writing style which is both accessible and concise.

—Rea Andrew Redd, Civil War Librarian

Written in popular fashion, [*Campaign for the Confederate Coast*] offers a rather unusually comprehensive summary of related topics and events that should appeal to a wide range of readers. Avid students of Civil War naval affairs will benefit from the collective reinforcement of their prior reading through Hahn's able synthesis, and those new to the subject matter are exposed to an introductory history with remarkable range.

—Andrew Wagonhoffer, Civil War Books and Authors Blog

Campaign for the Confederate Coast

Campaign for the Confederate Coast
by Gil Hahn

Copyright © 2021 by Gil Hahn

Published by
West 88th Street Press
14 Meadows Lane
Wilmington, Delaware 19807

All rights reserved. No part of this book may be reproduced or transmitted in any form or by any means, electronic or mechanical, including photocopying, recording, or by any storage or retrieval system, except in the case of brief quotations embedded in critical articles and reviews, without prior written permission by the publisher.

Cover art includes details from two drawings: *Flambeau* and *Patapsco*, Xanthus Russell Smith, Port Royal, South Carolina, United States, 1863-1927, Graphite and chalk on woven paper, 1967.2227, Bequest of Henry Francis du Pont, Courtesy of Winterthur Museum and *U.S. Naval Machine Shop, Port Royal S.C.*, Xanthus Russell Smith, Port Royal, South Carolina, United States, 1863, Graphite and chalk on woven paper, 1969.2726, Bequest of Henry Francis du Pont, Courtesy of Winterthur Museum

ISBN 13: 978-1-7349537-0-1 (print)
ISBN 13: 978-1-7349537-1-8 (eBook)

Library of Congress Control Number: 2020908772

Publisher's Cataloging-In-Publication Data
(Prepared by The Donohue Group, Inc.)

Names: Hahn, Gil, author.
Title: Campaign for the Confederate coast : blockading, blockade running and related endeavors during the American Civil War / Gil Hahn.
Description: Wilmington, Delaware : West 88th Street Press, [2021] | Includes bibliographical references.
Identifiers: ISBN 9781734953701 (print) | ISBN 9781734953718 (ebook)
Subjects: LCSH: United States--History--Civil War, 1861-1865--Blockades. | United States--History--Civil War, 1861-1865--Naval operations. | Confederate States of America--History, Naval. | Coasts--Confederate States of America--History--19th century.
Classification: LCC E600 .H34 2021 (print) | LCC E600 (ebook) | DDC 973.75--dc23

Content Editor: Gail M. Kearns
Copyeditor: Joni Wilson
Book and Cover Design: *the*BookDesigners
Book production coordinated by To Press & Beyond,
www.topressandbeyond.com

Printed in the United States of America.

CAMPAIGN *for the* CONFEDERATE COAST

BLOCKADING, BLOCKADE RUNNING *and* RELATED ENDEAVORS DURING *the* AMERICAN CIVIL WAR

GIL HAHN

West 88th Street Press
Wilmington, Delaware

Contents

1.
Aims and Means 1

2.
Evolving Tools of War 19

3.
Exceptional Commercial Circumstances 50

4.
Planning the Blockade and Blockading Tactics 75

5.
Mounting and Maintaining the Blockade 101

6.
Ramparts, Raiders and Rams 132

7.
Jealously Guarded Prerogatives 162

8.
The Campaign: 1861-1863 197

9.
The Campaign: 1864-1865 226

10.
Notes 256

1. Aims and Means

At the end of the first major battle of the Civil War, Confederate forces held the field near a creek called Bull Run while Federal forces withdrew toward Washington City, their retreat becoming ever more disordered until it resembled a rout. The victory in the first major battle elated the Confederates but dashed the hope many held that the war would be short and decisive. Indeed, the historical record contains many statements by people on both sides lauding their own side's martial abilities, patriotic ardor and the rightness of their political point of view and denigrating those of the opposing side, culminating in the belief that a single, major battle would end the war and resolve the political disputes that caused it. Historical experience tells us that such beliefs were not well informed, and, even if genuinely held, we must regard them as the bravado of people in a state of pitched political excitement. In the Confederate states, the excitement of the highly contested 1860 presidential election did not abate after the vote but continued through the creation of the secession conventions, the election of delegates, the conventions' decisions to secede, the formation of a Confederate national government and preparation for its defense by arms. In the states that remained committed to the Union, an extended period of uncertainty and confusion followed the presidential vote, but the flame of patriotic determination blazed bright again with the news that Confederate forces had fired upon Union-held Fort Sumter. Their beliefs, accordingly, were based upon their hopes and their feelings of righteousness rather than clear-eyed observations and rational analysis, and they failed to ask themselves the question: what would we do if we lost the first great battle but had the means to continue the war? That, in the first instance, should have indicated how the other side would behave if the circumstances were reversed.

In the aftermath of Bull Run, the *New-York Daily Tribune* observed, under the headline "Disasters on the Road to Victory," that "panic, flight, disaster, and a certain proportion of cowards, are to be looked for in all armies and all wars, and that they furnish no presumption at all unfavorable to ultimate success." Suggesting that the Federals might obtain "ultimate success" at that time reflected extraordinary optimism. The Confederate states, having seceded from the Federal Union, formed their own national government and an Army to sustain their claim of independence. From the firing on Fort Sumter to the victory at Bull Run, they showed both the intention and the apparent capacity to remain independent.

In the context of the Civil War, "ultimate success" meant something different to the Federals and the Confederates. For the Federals, it meant extinguishing the Confederate claim of independence and restoring the authority of the national government formed under the Constitution. Confederate independence depended upon the continued existence of the Confederate Armies, so a Federal victory depended on the capture or destruction of every sizable element of the Confederate Armies. For the Confederates, by contrast, victory meant survival rather than conquest—any outcome of the war that left Confederate Armies in existence and capable of sustaining Confederate independence was a Confederate victory.

At the time of President Lincoln's inauguration, about a month before the war began, Brevet Lieutenant General Winfield Scott, general-in-chief of the United States Army, presented four alternatives to the government. First, he suggested the adoption of proposed constitutional amendments, known collectively as the Crittenden Compromise, that purported to protect the legal status of slavery. Second, the government could collect tariffs offshore from the ports of the seceded states or close the ports and blockade them. Third, he proposed invading and conquering the seceded states, which he saw taking several years at great cost in human life. He proposed that the fruits of military victory, if it could be achieved, would be bitter and possibly fatal for the American tradition of liberty: the prize, as he saw it, would be "fifteen devastated *provinces*"—the emphasis was the general's to distinguish them from co-equal states—that would be subject to military occupation for generations, a circumstance that could lead, in his view, to the eclipse of American democracy with the rise of an emperor or a protector. Fourth, General Scott suggested, "Say to the seceded sisters States—*wayward sisters, depart in peace!*"

None of the suggestions could have been appealing to the Lincoln government, which desired to curb the influence of slave power generally and to preserve the Union, but the range of alternatives presented reflected the lack of a general consensus that existed in the non-seceding states during the period between the onset of the secession crisis and the firing on Fort Sumter. Indeed, General Scott's fourth suggestion echoed remarks made in Horace Greeley's *New-York Daily Tribune* just days after the 1860 elections:

> And now, if the Cotton States consider the value of the Union debatable, we maintain their perfect right to discuss it. Nay: we hold, with Jefferson, to the inalienable right of communities to alter or abolish forms of government that have become oppressive or injurious; and, if the Cotton States shall decide that they can do better out of the Union than in it, we insist on letting them go in peace. The right to secede may be a revolutionary one, but it exists nevertheless; and we do not see how one party can have a right to do what another party has a right to prevent. . . . We hope never to live in a republic, whereof one section is pinned to the residue by bayonets.

The *Tribune* insisted, however, that the exercise of secession had to be considered deliberately and approved by a general vote, neither of which occurred in the wave of secessions that swept through the cotton states. Such niceties became irrelevant when Confederate forces fired upon Fort Sumter, which shifted the mood of the North generally to militant defense of the Union.

After the war had begun, but prior to the debacle at Bull Run, General Scott outlined a strategy "to envelope the insurgent States and bring them to terms with less bloodshed than any other plans." The strategy, which became known to contemporaries and history as the Anaconda Plan, consisted of a "strict blockade" of the Atlantic and Gulf Coasts coupled with a drive down the Mississippi River to control its length to the sea but otherwise did not contemplate an invasion of the Confederate territory. Another Federal officer recalled General Scott explaining the rationale of his plan as follows: "you will thus cut off the luxuries to which the people are accustomed; and when they feel this pressure, not having been exasperated by attacks made on them within their respective states, the Union spirit will reassert itself," and he predicted that within a year "all difficulties will be settled." Invade the South at any point, he continued, and in a year's time "you will be further from a settlement than you are now." If we accept that General Scott said something like this, several fundamental flaws prevent accepting General Scott's strategy as a cogent military analysis. Not the least of these was that, given the existing technologies—and the uses being made of it—a "strict blockade" could not shut off the access to the world that the Confederates needed to sustain their Armies in defense of their independence without the further step of capturing the principal Confederate ports. Although this fact might have been suspected when the war began, in fairness it did not become known with certainty until later.

Immediately after the attack on Fort Sumter, the Federal government declared a blockade of the Confederate coast, and the Federal Navy Department began taking steps to build a Navy of sufficient size and strength both to enforce the blockade and to assist with the prosecution of the war along the coast and upon inland waters. Just after the Federal defeat at Bull Run made a longer war a near certainty, President Lincoln set down on paper a list of initial thoughts about waging war over the longer term. He began, "Let the plan for making the Blockade effective be pushed forward with all possible despatch."

The Federal blockade was not sufficient to achieve victory—it could not capture or destroy the Confederate Armies—but it was a necessary ingredient of the eventual Federal victory. Then, as now, war was an industrial undertaking, requiring vast quantities of supplies and weapons produced or collected and moved great distances to maintain large armies in the field. The Confederacy was largely an agrarian nation that produced economic wealth but lacked the capacity to produce much of the materiel its Armies required or to keep in repair the means of transportation needed to move it. By the same token, maintaining commercial connection with international markets of the world was not sufficient to assure a Confederate victory, but it was necessary to sustain Confederate Armies in active operations. The blockade impaired the Confederates' ability to convert their domestic economic resources into imported materiel and supplies to sustain their war and their domestic economy. If the Federals failed to mount the blockade or had maintained it less rigorously, the Confederates might have become more capable of resisting the Federal invasions, and the Confederate population generally might have been less burdened by the hardships of war, which could have prolonged a military stalemate and led to a negotiated peace—meaning a Confederate victory.

The ultimate Federal victory resulted from the war on land: the constant growth of Federal forces, the progressive weakening of the Confederate military and economy, the elevation of Ulysses S. Grant to command of the Federal Armies, and the implementation of his strategy of constant engagement. The success of General Sherman's Atlanta campaign contributed to President Lincoln's reelection in 1864, which assured that the Federal prosecution of the war would continue beyond the end of his first term of office in March 1865 and thereby dashed Confederate hopes of a negotiated peace with a post-Lincoln Federal government.

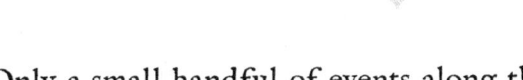

Only a small handful of events along the Confederate coast rise to general notice in the standard narratives of the Civil War. These include the captures of Port Royal, Fort Pulaski and New Orleans; the first battle of the ironclads at Hampton Roads; the failed ironclad attack on Charleston; and the captures of Fort Wagner, Mobile Bay and Fort Fisher. Although dramatic, these events shed little light upon blockading, blockade running, and the actions taken by the Confederates to defend their major ports and encourage a continuation of their international commerce. Monographs and unit histories illuminate limited aspects of the war along the Confederate coast, but they do not convey the complexity of the extended struggle on the one hand to maintain—and on the other hand to foreclose—the Confederates' access to the world. To understand the conduct of the Civil War along the Confederate coast, we need to examine the technology then available as well as the intentions, resources and limitations of each of the major participants—the Federals and the Confederates, who were the combatants; the neutral nations, primarily the British but also the French, who sought to advance their national interests; and nationals and monied interests of the neutral nations that sought to profit from the extraordinary economic situation caused by the Civil War. The purpose is not to examine complexity for its own value but rather understand the factors that made up the extended coexisting combat and commercial environment. As is always the case in war, the actions of each participant—belligerent and neutral alike—shaped or had the potential to shape the environment in which all of them contended.

The proposition that the blockade was essential to the eventual Federal victory rests upon the observation that the Confederates lacked the means to convert their considerable domestic resources into military and economic power; if the Confederates had been able to do so, the blockade would have been irrelevant. To appreciate the Confederates' potential vulnerability to a blockade, we need to understand the resources they had on hand at the start of the war, the productive potential of their agriculture and industry and the means available to access the world market.

That the Federals had an advantage over the Confederates in population and industrial capacity is true beyond question.

The total population of the contiguous states in 1860 totaled 30.7 million people of whom 62.8 percent lived in the Federal states and 37.2 percent lived in the Confederate states—roughly a third of the Confederate population were slaves. White men of "military age"—between the ages of 18 and 45 years—would form the Armies, and in 1860 the Federal states contained 3.9 million of these men (72.4 percent of the total), and the Confederate states contained 1.5 million (27.6 percent). As the Civil War began in 1861, neither Army accepted Black men into the ranks, although both employed Black people as noncombatant laborers. In 1863, after the issuance of the final Emancipation Proclamation, the Federals began accepting Black men into the Army, which increased their further manpower advantage. The Confederates did not begin discussing the possibility of using Black men as soldiers until just before the end of the war.

The 1860 census enumerated as "manufactures" the physical products of businesses that made sales of $500 or more during the year ended June 30. The aggregate value of the manufactures of the contiguous states was $30.7 million, of which 91.4 percent was produced in Federal states, and 8.6 percent was produced in Confederate states. When set against the benchmark of the respective shares of the total population (the Federals 62.8 percent and the Confederates 37.2 percent), this data shows that the Confederates had a substantially smaller manufacturing capacity than did the Federals. While not all manufacturing capacity could be converted to war production, and while both the Federals and the Confederates relied upon imports to equip their Armies, the Confederacy was at a clear disadvantage.

While the military advantage implied by the magnitude of these differences in population and manufacturing capacity seems overwhelming, it diminishes in light of the different Federal and Confederate conditions of victory. The Confederates could not remedy the disparity in population. By the battle of Bull Run, both Armies were roughly the same size. Afterwards, both Armies grew, although the Federal Army grew at a faster rate until it was significantly larger. Yet for much of the war, the Federals were unable to capitalize on their larger size, and this failure confirmed the view of many informed contemporary observers that the Federals could not restore the Union. Unlike the difference in population, the Confederates could address the disparity in industrial capacity. Before the war, the southern states had obtained most of their goods from the northern states, both domestic manufactures and imports. The war cut off this commerce (apart from smuggling, which persisted throughout the war), but the Confederates could continue to import goods from overseas so long as they had the ability to pay and their ports remained open. Although the Federal Armies were better supplied, the Confederates generally had adequate amounts of arms and munitions and were said never to have lost a battle due to lack of supply. (By comparison, the Confederate commissary performed poorly throughout the war, which increased the hardships felt in the Confederate ranks, especially in the east.)

Several important crops grew predominantly or exclusively in southern states, cotton being the foremost among these. During the 18th century, Britain developed machines for spinning cotton fibers into thread and weaving cotton cloth, which reduced the cost of cotton cloth, thereby increasing both consumer demand for cloth and industrial demand for raw cotton. Britain was the world's largest supplier of cotton cloth, and in 1860 about 80 percent of the raw cotton imported by Britain came from the United States, essentially all of it grown in the Confederate states. Cotton represented about half of the value of all exports from the United States. Cotton and the other principal crops of the Confederate states—rice, sugar, tobacco and naval stores (products made from pine sap)—provided the means to purchase the arms, munitions and other supplies that the Confederates needed to wage their war. Thus, the war in southern coastal waters determined the extent to which the Confederates could finance the war and acquire the arms and goods they needed to fight.

Only in the creation of a Navy and ironclad vessels did the Federals' greater industrial capacity translate itself into a clear military advantage. The Confederates never succeeded in offsetting the Federal naval power with their own Navy or alternative weapons, and Confederate shore defenses seldom withstood the force that the Federals could bring to bear.

The combination of the war, the Federals' declared blockade of the Confederate states and the Confederates' near monopoly in cotton fiber created profit opportunities for those with the capital and the daring to seize them. First, the blockade threatened to curtail the supply of American cotton to European mills, which increased the price of raw cotton in Europe. Second, the Confederates needed munitions to wage their war and had only a limited capacity to produce them domestically. Accordingly, so long as the blockade remained in place, munitions could be expected to command a premium price in the Confederacy. Third, aside from food, the Confederate states produced only a small portion of the manufactured goods they consumed. The war and the blockade threatened to interfere with importing goods into the Confederacy, which meant that the prices of all imports would rise and increase the profit of the sellers who brought them through the blockade.

Smuggling or trading through the lines that persisted throughout the Civil War offset to some extent the effects of the blockade. Although trading through the lines remained controversial, President Lincoln and some other highly placed persons in the Federal government condoned it. Northern goods nominally sold for export to neutral nations also reached the Confederates through the blockade, and while Federal officials took some steps to curb the more obvious abuses, such actions were not always effective.

In theory, a blockade runner could anchor at any point off the Confederate coast and put its cargo ashore in small boats. The weather or a Federal blockader could interrupt landing the cargo. Unless a railroad connection was near at hand, a cargo landed at a remote point lost much of its value as a result of the expense of transporting it overland to the place where it was wanted. Moreover, the wartime demand for manpower, iron and other resources crippled the Confederate ability to construct additional miles of railroad track, and the wear upon railroad track and equipment, coupled with the inability to procure repairs or replacement, reduced both the number of railroads that remained in operation and the carrying capacity as well as the speed of those railroads that continued to function.

AIMS AND MEANS

The Confederate coast—which was over 3,500 miles long and possessed 189 harbors, inlets and bays—appeared to provide many potential destinations for blockade running. As a practical matter, only a handful of locations satisfied the several requirements for being a suitable destination for blockade running in the 1860s. First, the destination needed to be a harbor that was protected from the weather and waters of the ocean. Second, the harbor needed to be of sufficient size and depth to accommodate the number and types of vessels that ran the blockade. A smaller and shallower harbor limited the number and size of vessels capable of using it. Third, the harbor needed to be defended against Federal attacks from water and land. As a part of its defense plans in the decades before the Civil War, the United States government had built stone and masonry forts to protect its principal harbors. The Confederates took over most of these forts in their states and built additional defenses as their resources permitted. Fourth, it needed wharves and warehouses to facilitate the loading, unloading and transfer of cargoes. Fifth, it needed transportation to points inland to move cargoes to and from the harbor. Transport between points on the coast was less expensive by water, and most of it was powered by wind rather than steam, but the blockade suppressed most of this coasting traffic. Railroads were the most efficient means of overland transport in the 1860s, but, over any distance away from the railroad, horse-drawn wagons were the principal alternative.

In addition, the harbor needed a sufficiently large community of pilots to navigate arriving and departing vessels. Southern ports tended to be shallow, and the sandbar erected and maintained by tidal action before the entrance to most harbors—referred to as "the bar"—made access more difficult, although channels through the bar enabled deep draft ships to pass. The deepest channel into Charleston was the Main Ship Channel that was 12 feet deep at the ebb tide and 17 to 18 feet deep at the flood tide—it had been dredged in the years just before the war. The locations of channels tended to be stable, although they could shift over time, but the banks and shoals along the channels were more changeable and made a pilot's intimate knowledge of the channels valuable. Ships bound for Charleston regularly took on a licensed pilot before crossing the bar as a matter of prudence. By contrast, the water over the bar before New York City was 21 feet deep at ebb tide, but the law required larger vessels entering and leaving New York to have pilots.

The causation was circular: aside from the physical attributes of the place, the importance of certain harbors created the need for amenities that made it an attractive destination—such as defenses, wharves, warehouses and pilots—and the existence of these amenities made the harbor a more attractive destination. During the war, only two of the lesser southern harbors saw a substantial amount of blockade running. Import data for 1860 suggest that southern harbors were insubstantial. New York City had the largest share of imports with 64.5 percent of the total, and New Orleans was a distant second with only 6.3 percent. Export data indicate that the southern ports varied but were active. As Table 1-1 shows, New York City had the largest share of exports with New Orleans a close second. The next three largest were also southern. Together those four southern ports accounted for about half the total value of exports from the United States in 1860. About 90 percent or more of the value of the 1860 exports from each of those four southern ports consisted of cotton, and they shipped about 90 percent of the cotton exported from the United States. Table 1-2 shows the amount of cotton that passed through the four main southern ports during 1860. The numbers indicate that each of these places possessed a substantial capacity to handle cargoes.

TABLE 1-1: PERCENT OF EXPORTS	
New York City	32.2%
New Orleans	28.9%
Mobile	10.4%
Charleston	5.7%
Savannah	4.9%
Boston	3.6%
Source: Report of the Secretary of the Treasury, *Transmitting a Report from the Register of the Treasury for Commerce and Navigation for the Year Ending June 30, 1860* at 350-351.	

TABLE 1-2: 1860 COTTON EXPORTS IN MILLIONS OF POUNDS	
New Orleans	1,835
Mobile	655
Charleston	345
Savannah	300
Source: Report of the Secretary of the Treasury, *Transmitting a Report from the Register of the Treasury for Commerce and Navigation for the Year Ending June 30, 1860* at 316-317.	

Peacetime patterns of trade determined the major cotton ports in the states that would form the Confederacy, but the pressures of war closed ports or diverted blockade runners to less well-defended destinations, thus raising the importance of other ports that had enjoyed less traffic before the war. These included Wilmington, North Carolina; St. Marks, Florida; and Galveston, Texas.

Principal and secondary harbors may not have had a monopoly on the blockade-running traffic. Blockade-running vessels may have gone to smaller, seldom-used harbors to pick up a cargo of cotton, either from the stores of a nearby cotton plantation or a supply assembled for the purpose of a transaction. Smaller vessels, including some purpose-built steamers with shallow drafts, were suited to such voyages. Unlike a visit to a well-trafficked port where the blockade runner would expect to find cotton, a visit to a less well-trafficked destination would be worthwhile only with the knowledge that an ample supply of cotton would be available there, which required that the fact be communicated in advance. The isolated nature of these less well-trafficked destinations would have made them more attractive to blockade runners because of the reduced Federal presence—with most of the blockaders at the major ports, perhaps only an occasional patrol would pass by. Also, later in the war the Confederate government passed laws and promulgated regulations to restrict inbound traffic to necessities, to reserve a portion of the inbound and outbound cargoes for government use, and to bind the blockade runners economically to keep them in the blockade running business. Under these circumstances, blockade runners might choose to use a less-trafficked destination to trade more freely and reap fuller profits. In an isolated backwater, no government agent would be around to enforce the law.

Just as a well-defended harbor provided a place of refuge for the blockade runner, so too did international borders: the rules of international law prohibited a blockader from stopping or capturing a blockade runner in the territorial waters of a neutral nation. The foreign possessions off the American coast provided sanctuaries for vessels preparing to run the blockade. Mexico—especially Matamoras, across the Rio Grande from Brownsville, Texas—provided a putatively neutral destination that was relatively secure from interference by Federal blockaders so long as the Confederates did not own the inbound and outbound cargoes. And the trade flourished accordingly: Matamoras lacked a harbor, but cargoes moved between ship and shore on lighters. In 1863 the commander of the West Gulf Blockading Squadron reported that, whereas half a dozen vessels might have called at Matamoras during the year before the war, at present 180 to 200 vessels were waiting to discharge or take on cargo.

Mexico was not an anomaly in the Civil War but a symptom of what made the war on land different from the war in the littoral and the oceans beyond. With the exception of Mexico, foreign nations did not figure directly in the calculations of any field commander, whereas foreign nations and foreign nationals figured large in the calculations of most every Federal officer serving on coastal waters and beyond. They encountered foreign warships, merchant vessels flying foreign colors, and foreign nationals attempting to conduct commerce with the Confederates in spite of the Federal blockade. An encounter with any of these could run afoul of the rights and obligations accorded to belligerents and neutrals and thereby had the potential to escalate into hostilities or a foreign war that would be detrimental to the Federal effort in the Civil War. (Mexico was different because it was an increasingly important source of supply to the Confederates west of the Mississippi, which Federal land forces took steps to interdict. Federal commanders in Texas were also mindful of the French forces that had invaded Mexico in 1862, overthrown the constitutional republic and supported the regime of an emperor.)

One such hostile contact occurred when the Federal cruiser *San Jacinto* stopped the British mail packet *Trent* and arrested two Confederate diplomats on their way to Europe. Britain threatened war if the diplomats were not released, and the Lincoln government relented. War tensions between Britain and the United States also rose as the Confederates contracted with private shipyards in Britain to build vessels to be used as Confederate cruisers that captured and destroyed a number of Federal merchant vessels. Canada at the time consisted of British colonies, and several incidents involving Canadian territorial waters and the Canadian border threatened war between Britain and the United States—for example, one group of Confederate sympathizers conducted a raid from Canada into St. Albans, Vermont, while another group hijacked a commercial vessel in or near Federal coastal waters and sailed it to Pubnico, Nova Scotia, where, on inland Canadian waters, pursuing Federal cruisers had a tense confrontation with British authorities. And, as mentioned above, the distraction to both Federals and Confederates caused by the Civil War created an opportunity for France to involve itself in the affairs of Mexico. The international was a volatile element that infused all of the campaign for the Confederate coast.

2. Evolving Tools of War

The tools of war undergo periodic change. New developments initially supplement technologies that are in use and replace them only eventually. As a result, any consideration of what technologies existed at a given point in time necessarily involves a look at what was already in use and what had just come into being. In terms of the war along the Confederate coast, for example, in addition to thinking about the ironclad, turreted steam-powered vessels like the *Monitor* and her numerous sisters, we need to be aware that wooden ships, some driven by steam and some driven only by the wind, also remained in use. The technological changes—some decades old and some very recent—that worked themselves upon warships made them a world apart from the vessels that had dominated the age of fighting sail. By contrast, although some technological changes had the potential of altering the coastal combat environment, with one exception, it remained largely unchanged. The exception, the use of spotlights for nighttime operations, was not carried to its logical extension in the blockade itself. Inasmuch as early-21st-century littoral combat environment is distinctly different from what it was during the Civil War, a proper appreciation of the past requires that we examine the conditions in the mid-19th century.

The ships that fought in the War of 1812 were of many sizes and shapes, but they all shared three features with one another and with warships that had been in use for several hundred years: they were built of wood, their armaments were smoothbore cannon, and they were propelled by the wind. Human muscle was the principal source of power used to perform work on board these vessels—to raise the anchor, trim the sails and aim the cannon. Fire was used for limited tasks aboard warships—cooking, illumination, shaping iron and discharging the cannon—and although by 1812 steam had been used to propel some civilian vessels, steam was not used aboard warships.

Technological change brought an end of the age of fighting sail. Steam propulsion, more powerful artillery and iron armor (and iron hulls) did not emerge as fully developed technologies but rather as novel ideas or applications of existing ideas that opened extended periods of experimentation and development. Some new technologies impose immediate effects on the people who use them—some distinctly different equipment and skills are needed to operate a sailing ship as compared to those needed aboard a steamship. Other effects of technological change are slower in coming inasmuch as the full implications of a change may not be recognized until the technology has been in use for some time. Preexisting "conventional wisdom" may limit the uses to which the technology is put. Pure insight might provide the inspiration that makes fuller use of the changed technology, but a flash of insight cannot be expected to occur, and when it does occur the person possessing the insight might not command the wherewithal to effect the changed usage. Changed circumstances, experiment or accident might be needed to provoke the insight to overcome the inertia of conventional wisdom or to inspire the new use to which the technology might be put.

Steam made a vessel largely independent of the wind and the tide, and it exchanged dependence upon the wind and sails for a dependence upon machinery and the supply of fuel. Although steam made warships more maneuverable and faster, it also increased their vulnerability: a shot that pierced the boiler immobilized the vessel, and the explosive release of steam endangered the crew. Through the 1850s the United States Navy had continued to build wooden-hulled warships and fitted them with steam engines. Line officers regarded steam power as auxiliary, but engineering officers believed that the Navy eventually would come to rely solely on steam. Given the state of the steam engine's development, the civilian shipping industry agreed with the line officers. Although several shipping companies operated steam-powered vessels on cross-ocean routes, the larger part of the world's ocean traffic still moved under sail.

Notwithstanding these prevailing opinions, the certainty that each of the Federals, the Confederates and the blockade runners to some extent would employ steam-powered vessels, encouraged the others to resort to steam. And as a result, most of the actors in the blockade conflict incurred various difficulties related directly to the state of development of maritime propulsion as it existed in the early 1860s. Most but not all. Sailing ships continued to challenge the blockade throughout the war, and the Federals, despite the avowed preference for steam stated by the Secretary of the Navy, continued to acquire and employ sailing ships if only because they could float, carry guns and thereby increase the number of blockading vessels.

The Navy's engineers, perhaps foreseeing the need to keep boilers alight and engines operational for months on end, preferred engine designs that were simple and durable rather than more sophisticated designs that, while more efficient in the use of fuel and steam, could be more prone to break down. The wartime expansion of the Federal Navy fleet created a demand for engineers, and the Navy could not find enough men with adequate training and experience. The simpler and more robust engines used in Navy vessels were more in line with their operators' capabilities. The potential rewards from blockade running made it less difficult attracting qualified engineers who were subjects of neutral nations. Finding qualified personnel was more difficult for Confederate vessels launched from Confederate shores. The Confederate states had fewer qualified engineers to begin with, and the needs of the Confederate Army swept into the ranks a substantial number of those; the Army was hesitant to release any man from service for any reason. Confederate cruisers launched from foreign shores were better able to find qualified men. The cruisers sailed with Confederate officers, and the crews who signed on generally were subjects of neutral nations and were attracted by the expectation of economic reward.

Other factors tended to affect Federal, Confederate and blockade-running vessels alike. Operating a boiler, for example, consumed steam and required adding water to the boiler from time to time. Minerals in the water encrusted the interior of boilers with a solid, called scale, and eventually hampered the transmission of heat and thus the generation of steam. Charging the boiler with seawater worsened the problem. Many vessels were equipped with a condenser that produced fresh water for both the boiler and the crew. Having a condenser did away with the need to keep a supply of water in barrels, but operating the condenser increased the rate at which the fuel supply was consumed.

Steam engines in the Civil War era generally were designed and built one at a time. Although some products, like rifles and clocks, were mass produced with interchangeable parts, the techniques of mass production had not been extended to steam engines, which made them costly and time consuming to build and to make replacements for parts damaged or worn out.

The earliest steamboats used paddlewheels to impart motion, and a successful prototype propeller was developed in 1827. Blockade runners preferred paddlewheels: of the 353 steamships associated with blockade running, about 67 percent used paddlewheels, 19 percent used a single propeller and 6 percent used twin propellers. The paddlewheel and the propeller each imposed different requirements upon the design of the ship and its machinery, some pertaining to all vessels, and some particular to armed vessels. First, the paddlewheel was most efficient with only the bottom sixth submerged—if deeper, the engine labored harder, and, if shallower, the wheel slapped the water. Consuming coal lightened the ship, causing it to ride higher in the water, which affected the efficiency of the paddlewheels. Rolling seas would also plunge the paddlewheels into and out of the water. By contrast, for the propeller to function properly, it had to be entirely submerged, and the depth made little difference, although a rolling sea could lift the propeller out of the water as well.

Second, since the driveshaft that turned the paddlewheels was above the water, the mechanism did not compromise the integrity of the hull. On the other hand, the driveshaft of a propeller ship passed through the hull underwater, which made leakage more of a concern.

Third, for military purposes, the location of the propeller underwater and the drive mechanism deep in the hull made them less vulnerable to enemy fire. Paddlewheels, by contrast, presented an obvious target for enemy fire. A paddlewheel warship mounted its guns fore and aft where the wheels did not restrict the field of fire.

Fourth, for a propeller to accomplish the same work as paddlewheels, the propeller needed to spin faster than the paddlewheels. This required either running the engine at higher speed or using additional gearing to increase the rate of rotation of the drive shaft.

The most profound effect of steam propulsion was on maneuver. In the War of 1812 the USS *Constitution* and the HMS *Guerrière* fought an extended battle while sailing roughly parallel courses, each turning to gain advantage and to avoid exposing her bow or stern to the other's broadside. Over the course of an hour, their travel remained a straight line. During the Civil War, the USS *Kearsarge* encountered the CSS *Alabama* off Cherbourg. Both steam-powered vessels steered circular courses—impossible under sail—as they fired at each other, completing about five full revolutions in an hour before a shell from the *Kearsarge* struck the *Alabama* at the waterline, sinking her.

Speed and the capacity for constant motion were also primary benefits of steam. The top speed of the faster Civil War era steamers was around 17 knots, slightly in excess of the faster sailing ships whose speed at any given moment was tied to the wind.

These benefits did not come without a price. Accidental boiler explosions occurred on both Federal and Confederate vessels that served during the Civil War. The boiler on the USS *Chenango* exploded while she was cruising out of New York harbor on April 14, 1864, scalding 35 of her crew, of whom 28 died. A coroner's inquest was convened to determine the cause of the explosion; the evidence pointed to several possibilities including faulty design, improper construction and low water. A warship's boiler was burst by an enemy shell in a number of Civil War actions. Side-wheel steamers were most vulnerable to damage because their boilers and steam engines generally sat higher in the structure of the vessel. Although the boilers and steam engines of screw steamers sat lower in the structure of the vessel, generally below the water line, they also suffered boiler damage from enemy fire.

In the matter of hull construction, the blockaders and the blockade runners had reached a technological transition, but differing economic conditions on the two sides of the Atlantic Ocean produced an equilibrium in which the old and new coexisted. Ships' hulls were traditionally made of wood, and in America they continued to be made of wood, which was plentiful and cheap. In Britain, by contrast, wood had been becoming increasingly scarce and, accordingly, increasingly costly. Supplied by a well-developed iron industry, Britain's shipbuilders had become accustomed to building ships with iron hulls, and many purpose-built blockade runners built in Britain had iron hulls, and a few had steel hulls. (Britain continued to build some wooden-hulled vessels. The cruiser *Alabama*, built in Britain for the Confederates, was of wooden construction.) Iron hulls possessed several advantages over wooden hulls: they were more buoyant; they possessed a potentially longer economic life in that they did not become waterlogged; they permitted construction of larger-sized vessels; and they endured better the constant vibration of steam engines mounted in them. Most of these advantages were not of assistance to a blockade runner in the Civil War inasmuch as the war did not last long enough for the iron hulls' endurance to pay economic dividends.

The increased destructiveness of artillery resulted from three developments: the replacement of solid shot with exploding shells, the increased size of the projectiles fired and the introduction of rifling into artillery.

Artillery had fired exploding projectiles throughout the age of sail, but ships firing upon one another, and shore batteries firing upon ships, generally fired solid metal projectiles—shot to pierce the hull, case and canister against personnel, and bar and chain shot to damage sails and rigging. Starting in the 1820s a French naval officer advocated the use of exploding shells against naval targets—ships of the time were constructed of wood, cloth and fiber soaked in tar (rendered pine sap), which made them vulnerable to explosion and fire. In the battle of Sinope during the Crimean War (1853-1856) a Russian fleet firing shell guns destroyed all but one of the vessels of the Turkish fleet.

In the pre-Civil War decades, the United States Army and Navy also began making larger guns to fire larger projectiles, propelled by larger amounts of gunpowder. The larger guns were made from cast iron, which was relatively inexpensive, plentiful and easy to cast into large pieces, but they were more prone to bursting. Ordnance officers tried to address this problem by improving the quality of the metal used, revising the design of the guns, and altering gunpowder to change the rate at which it burned.

A further development that added power to artillery was rifling—spiral grooves on the inside of a barrel that engaged the projectile and imparted a spin to it that enabled the gun to throw projectiles more accurately and with greater force. Even though the largest rifled cannon in use during the Civil War were smaller than the largest smoothbore cannon, the greater pressures that developed within the rifled guns increased the danger of bursting. Some manufacturers addressed the problem by wrapping the breech with bands of wrought iron, such as the Parrott rifles used by the Federals and the Brooke rifles used by the Confederates. Both rifled cannon and smoothbore cannon burst during combat, but rifled cannon burst more often.

The increased destructiveness of naval artillery in general, and when directed against wooden ships in particular, led to the third principal change from the warships: metal armor to increase their defensive power. In the late 1850s both France and Britain built experimental armored ships. Ship designers experimented with different types of iron and various types of iron plate construction. By 1862 when ironclad vessels first fought one another in the American Civil War, other nations of the world had built, were building or were planning to build at least 88 warships of various sizes with iron armor.

The pendulum swing toward defense caused by armor led ordnance officers to improve their weapons. The lessons that the Federals drew from the first inconclusive battle between the USS *Monitor* and the CSS *Virginia* (formerly the USS *Merrimack*) were the need for both bigger guns and larger gunpowder charges. The *Monitor*'s 11-inch smoothbore cannon fired a 166-pound shot with a standard 15-pound gunpowder charge. The Federals produced a new 15-inch smoothbore gun that fired a 440-pound shot. The Navy's Ordnance Instructions at the time of the *Monitor-Virginia* fight said that 15 pounds was the standard gunpowder charge used in the 11-inch gun under all conditions. This gave the gun acceptable range and striking power and yet afforded it an impressive useful life. Subsequent experiments showed that occasionally firing an 11-inch gun with a charge of 20 or even 30 pounds did not greatly shorten its useful life, and larger charges were authorized when the circumstances warranted.

Rifled cannon firing solid metal bolts presented another challenge to armor. The closer tolerances required for a rifle converted more of the gunpowder's energy into thrust, and the end-forward flight of the bolt encountered less resistance passing through the air than a spherical shot of equal weight. Nonetheless, a number of naval officers believed that their larger smoothbore guns were more effective against armor than rifled cannon when fired at close range.

Another approach was to use hardened or denser metal projectiles. Most cannonballs and shells used during the Civil War were made from cast iron. Such projectiles sometimes shattered when striking armor. Heating the metal until it glowed and then cooling it quickly hardened the metal; such "chilled shot" was less likely to shatter upon impact. While preparing to attack the Confederate ironclad *Tennessee* in Mobile Bay, Federal Admiral David Farragut read about the effectiveness of steel shot against iron armor in *The Scientific American,* and he asked the Navy Department to provide some, which it apparently did. Similarly, the Confederates also acquired some wrought iron shot but lamented not getting steel shot.

The Federal Navy initially took a very different position with respect to heavy projectiles in general and wrought iron shot in particular. John Ericsson, who designed and built the *Monitor,* knowing that the Confederates were building an ironclad, prepared a number of wrought iron cannonballs for her 11-inch smoothbore guns. These were substantially heavier than the standard-issue, cast iron cannonball, which led to the Federal Navy Bureau to forbid their use for fear that they would cause the guns to burst. When the CSS *Virginia* encountered the *Monitor* in Hampton Roads on March 9, 1862, she had expected to attack wooden warships, and she was armed with rifled cannon but carried shells and hotshot rather than solid metal bolts. Thus, neither ironclad was fully prepared to fight another ironclad.

From time to time a story that is so colorful or illustrative of a point enters the standard narrative of history, even though it is at odds with the established facts of the historical record. One such tale is that when General Irwin McDowell, in explaining his plan for the first Bull Run campaign to President Lincoln and the cabinet, remarked upon the inexperience of his troops, the president replied, "You are green, it is true; but they are green, also; you are green alike." It did not happen. The words quoted were spoken by General McDowell in his testimony before the Joint Committee on the Conduct of the War, and he attributed them to General in Chief Winfield Scott. In his testimony, General McDowell goes on to say that when he and General Scott went to the White House to explain the plan, the cabinet received it without comment.

The observation is relevant here because a similarly apocryphal story exists about the armament of the *Monitor*. A number of years before the Civil War, John Ericsson designed and built an innovative vessel for the Navy called the USS *Princeton*—the Navy's first screw steamer, all the Navy's other steamers being paddlewheel vessels. The *Princeton* carried a broadside of carronades and two large 12-inch cannon. Mr. Ericsson designed one of the large guns. Because of the great pressure that would be built up by its large projectile and powder charge, Mr. Ericsson reinforced the breech end of the cannon's cast iron barrel with bands of wrought iron as were the later Civil War era Brooke rifles and Parrott rifles. The other large cannon also had an enlarged breech section that mimicked the shape of Mr. Ericsson's cannon, but it was made entirely of cast iron. The different characteristics of wrought iron and cast iron meant that the breech reinforced by wrought iron was stronger than the breech made entirely of cast iron. As a result, during a demonstration of the *Princeton* and her armament, one of the large cannon (not the one designed by Mr. Ericsson) burst, killing six—including the secretary of state and secretary of the navy—and injuring about 20 others. The often-told story goes that when the *Monitor* was launched, the Navy, recalling the *Princeton* incident, decreed that the powder charge used in the *Monitor*'s guns should be reduced from 30 to 15 pounds. This reduction is offered as a partial explanation of why the *Monitor* failed to do more damage to the *Virginia*. As we have seen, the standard charge for the *Monitor*'s guns was 15 pounds, and the Navy did not authorize the use of a 30-pound charge until later in the war. (The story may have arisen during the Civil War. In testimony about the *Monitor-Virginia* fight before the Joint Committee on the Conduct of the War, a witness emphasized that 15 pounds was the standard charge permitted in the ordnance manual at the time.) Moreover, the story does not make sense. Ericsson designed and built the *Monitor*; the Navy provided the guns, so the Navy had no reason to be wary of Ericsson in respect to the guns. What is tantalizing is that Ericsson did provide wrought iron cannonballs for use in the *Monitor*, and the Navy forbade their use on the possibility that they would cause the guns to burst. Perhaps the factual cannonball story inspired the widespread but counterfactual gunpowder story.

While changing technology was having profound effects upon warships and their armament, the coastal combat environment was little changed. The existing coastal defenses had been planned during the age of fighting sail in light of the conditions that existed at that time. In spite of the technological changes that had come to pass, signaling between vessels was virtually unchanged from previous centuries. The means of detecting other vessels at sea were unchanged; the telltale signs that revealed the presence of another vessel at a distance did change—whereas sails billowing on a mast betrayed a sailing ship, a plume of smoke betrayed a steamer. The dark of night made detection all but impossible.

The various factors that made a port attractive as a destination for shipping did not include its military defenses, but the masonry and stone forts that defended the principal Confederate ports had a material influence on the coastal combat environment. Such forts protected all the principal harbors of the United States. The conventional military wisdom of the decades leading up to the Civil War was that forts generally could withstand an attack by ships. The ships' alternative tactic was to run past the fort, which reduced the time spent under fire but carried additional risks, such as becoming stranded while under the fort's guns and facing an enemy of unknown strength beyond the fort. On the other hand, forts were vulnerable to capture from land by a determined and well-equipped enemy, and the expectation was that a fort could resist capture for about two weeks. In the Civil War, steam power and rifled cannon upended the conventional wisdom. Steam-powered ships could run past forts with greater confidence; and a bombardment with rifled cannon could breach a fort's wall in days or hours rather than weeks.

Although the design of each fort that the United States built prior to the Civil War was specific to its site, each fort shared certain similarities. First, the forts were built of brick and stone. Earthen forts were more resilient and cheaper to build but expensive to maintain due to erosion. The small caretaker force manning a fort in peacetime was incapable of maintaining an earthen structure.

Second, many of the structures consisted of one or more tiers of casements behind an outer wall that could be more than 20 feet thick. Casements were open rectangular spaces covered over and separated from one another by masonry walls. Lined up like cells in honeycomb, each casement mounted a gun that fired through an opening in the outer wall. In forts built without casements, like Fort McHenry in Baltimore, guns were mounted on a single tier on top of the solid wall.

Third, the forts tended to be irregularly shaped, five-sided structures with sizes and shapes that varied according to defensible qualities of the terrain, the relative importance of the site and the capacity of the ground to support the structure. A fort that was vulnerable to land attack had rectangular or triangular works called bastions superimposed upon its corners. Fort Monroe in Virginia is an example of a fort with bastions. Guns on the walls could lay down fire along the adjacent walls, so an attacking force was vulnerable at all points of the fort's perimeter. Forts located where land attack was thought not possible, such as Fort Sumter on a manmade island in Charleston Harbor, generally lacked bastions.

Fourth, the forts were located near the water's edge, rather than being placed upon a hill or rise above the water. The shot and shell ricocheted across the water, like a stone skimmed over the surface of a pond. Shots from a higher elevation tended to plunge and were less likely to strike home.

Although originally intended to protect harbors and cities from hostile forces, during the Civil War the forts assisted blockade runners in entering and exiting Confederate ports, but some forts were better situated than others to fulfill this role. The forts protecting New Orleans sat 100 miles up the Mississippi River and thus did not prevent Federal vessels from entering the river and closing it to blockade running, although other water passages to the city remained open. Similarly, Fort Pulaski, which defended Savannah, was sited well inland. When the Confederates abandoned Tybee Island, where the Savannah River flowed into the sea, the Federals took control of the primary water route to Savannah and closed it to blockade running.

The Confederates placed obstructions—a sunken hulk and pilings—to prevent passage through Wall's Cut, a short canal that connected the Savannah River delta with a nearby navigable river. The Confederates did not establish a nearby presence to protect the obstructions, although their patrol boats passed nearby from time to time. Over the course of four days and nights, the Federals removed the obstructions, enabling their gunboats and other vessels to move upriver from Fort Pulaski without passing under its guns. The bank of the Savannah River was composed of several large marshy islands, and using materials and men brought in by boat through Wall's Cut, the Federals built a gun platform at the edge of the Savannah River constructed of wooden beams set upon a built-up foundation of sand and sandbags. Once that platform had been established and armed with cannon, the Federals established another battery just opposite it on a small island in the center of the river. While these batteries did not participate in the reduction of Fort Pulaski, they prevented communication by the Confederates with the fort from upstream by larger vessels. Smaller vessels could still reach the fort unseen through nearby creeks and marshes.

During the war against Mexico (1846-1848), the United States Navy engaged in cutting-out operations to seize or destroy enemy vessels in harbors and supported amphibious operations using ships' boats—oared vessels—to project power toward the shore. Cannon were not standard equipment on these boats, and they were armed with whatever smaller guns could be obtained. These improvisations served their intended purpose, but they also pointed up the need for a new weapon.

After the war, Lieutenant John A. Dahlgren designed several small cannon that could be mounted aboard boats and removed for use as field guns. Based on the Navy's experience, Lieutenant Dahlgren expected that these guns would be used against personnel rather than fortified positions or vessels, and thus he designed them to fire canister, shell and shrapnel rather than heavier solid shot. A modest powder charge gave these lighter projectiles a good range and permitted the barrel to be kept relatively thin at the breech, which reduced the weight of the weapon, making it easier to maneuver and transport. The barrel of a boat gun was shorter than a field gun of comparable caliber, which also reduced its weight. The shorter barrel compromised accuracy, but all the intended projectiles sent a spray of metal, which reduced the need for accuracy.

Lieutenant Dahlgren called his guns "boat howitzers." Two smaller boat howitzers were 12 pounders: one weighed 760 pounds and the other 430 pounds. By comparison, the 12-pounder Napoleon, a bronze field gun much used by Civil War armies, weighed about 1,230 pounds.

The carriage for mounting a howitzer aboard a boat consisted of two wooden boards clamped one flat on top of the other with a pair of bolts. The howitzer attached to the top board, and the friction between the two boards absorbed some of the recoil. The boat carriage rested on a small wooden platform in the bow and attached to one of three metal brackets mounted on the inside of the hull—one at the bow and the other two to the left and the right—and these permitted the gun to move in various directions over the bow. The howitzer also had a cast iron carriage that permitted its use ashore as a field gun.

Boat howitzers saw much use during the Civil War in boat expeditions or when guarding against hostile action—boats on nighttime picket duty had boat howitzers in addition to pistols, muskets and cutlasses. In preparing his squadron to run past Forts Jackson and St. Philip prior to the capture of New Orleans, Flag Officer Farragut had boat howitzers mounted on boat carriages in the masts and boat howitzers on deck mounted on field carriages. The monitors that served as picket boats in the blockade of Charleston late in the war were armed with boat howitzers on their decks. Other large vessels had boat howitzers on their decks to repel boarders and to support the attacks of their own boarding parties. Some smaller vessels, such as tugs, carried boat howitzers as their principal weapons.

The improvisation of the Mexican War became the standard issue of the Civil War, enabling boats and smaller vessels to become more formidable.

Communications at sea were limited by line-of-sight, and most of the means in use in 1860 were long established: signal flags. Ten flags with different colors and patterns represented the numbers zero through nine, and additional flags had specialized meanings, such as indicating that a message was being repeated or answering that a message had been received. Numbers from one to four digits in length represented different words and phrases. The 1861 signals book used by the United States Navy listed 8,889 words and phrases that permitted the reasonably efficient communication of complex messages. A simplified version of 33 signals enabled boats' crews to send messages in emergency situations. Vessels conveyed the same information at night by hoisting white, red and blue lanterns singly or in groups to represent digits. A rocket indicated that a message followed, and a lantern waved by hand indicated that the message was received. In 1859 the United States Navy approved the use of Coston signal flares. Coston signal flares burned as red, green and white lights, either singly or in sequence to represent the numbers zero through nine and the letters P for "preparatory," and A for "answering." Just like flags, the numbers indicated words and phrases in the signal book. Boats without Coston flares made these signals by combinations of rockets, lanterns, blue lights, flashes and discharges of boat howitzers and small arms.

Flashes were made with a flashpan, a shallow copper bowl, large enough to hold an ounce of gunpowder, held upright with a handle 2 feet long.

In terms of maintaining the blockade, the particular need was to alert other blockading vessels about the location and the direction of a blockade runner once contact had been made. Under the best of conditions—during daylight and in clear weather—signaling at sea was slow, and the reliable receipt of the intended message decreased as the distances grew larger. Blockade runners favored moonless nights and mist or rain for the obvious reason that they decreased their risk of detection. Even if the blockaders discovered a blockade runner, the signaling generally could not pass information quickly enough to effect a capture. Moreover, as we shall see, blockade runners used their own rockets and lanterns to confound the blockaders' efforts to communicate.

In 1858 a British Navy officer developed a signal lamp with shutters that could transmit Morse code, but it did not come into use until after the Civil War.

The flames of lamps and candles illuminated the world at night in 1860 as they had in 1812, although with some improvements. One standout development was the intense white light produced by burning a piece of calcium oxide (also known as lime) in a hydrogen-oxygen flame. Known variously as a Drummond light, a calcium light and a limelight, this instrument was a spotlight in theatres and illuminated evening ice skating at Central Park in New York City.

Calcium lights were not standard military equipment, but both the Confederates and the Federals found uses for them. The Confederates acquired a light in anticipation that the Federals would attempt to pass Forts Jackson and St. Philip on the Mississippi River in order to attack New Orleans, and they installed the light in Fort Jackson, but it was destroyed during the Federal bombardment of the forts. The Confederates ordered three calcium lights from Britain for the defense of Fort Fisher, but they got as far as Bermuda when the Federals captured the fort. The Federals made extensive use of lights in their night attacks on Confederate defenses in Charleston Harbor, and their use of lights was instrumental in the Confederate decision to abandon Fort Wagner. Later in the war, the Federals placed calcium lights on the decks of the monitors stationed in the ship channel at Charleston, but they were removed because the lights attracted fire from nearby Confederate forts. The Federals also used calcium lights to hamper Confederate operations on the James River in January 1865.

One curious omission is that the Federals did not employ calcium lights on any of the ships outside the blockaded harbors—beyond the range of the Confederate shore batteries—to observe the harbor inlet for outbound vessels and to scan the horizon for inbound vessels. Such brilliant searchlights would have all but eliminated the blockade runners' covering darkness and might have deterred most blockade running. Lieutenant Commander Stephen P. Quackenbush, in a letter to Secretary Welles in February 1863, recommended using calcium lights to detect blockade runners, but nothing seems to have come of it, perhaps because the suggestion came from a lieutenant commander rather than the squadron commander.

Admiral S.P. Lee requested calcium lights for the monitors to be assigned to the North Atlantic Blockading Squadron, and when Admiral Porter took over that squadron in late 1864, he issued a general order stating that calcium lights would be issued to the blockaders—"They will be found very useful in lighting up the bars and also while in chase." The Navy Department acknowledged and acted upon Admiral Porter's request shortly before the first Federal attack on Fort Fisher. With the capture of Fort Fisher, most of the blockade running came to an end, so the lights' usefulness as a deterrent to blockade running was not tested.

The practice of criticizing historical figures for the actions they failed to take is fair when those actions are within the common experience of the times, but it is unfair to lay such criticism for the failure to anticipate or invent the future. The failure to use searchlights across the water away from the shore to detect and deter blockade runners should be regarded as a missed opportunity. Another example we shall see was the failure of the blockading squadrons to realize that their vessels' night running lights, while deterring collisions, were assisting blockade runners by broadcasting the blockading vessels' positions while also providing an aid to local navigation.

The weapons and other tools useful in war undergo periodic testing in actual combat situations. While some of these are the major battles that attract most of the attention in military history, some smaller encounters may provide useful insights into the abilities, limitations and vulnerabilities of those weapons and tools. A remarkable series of encounters between small numbers of Federal Navy vessels and limited Confederate forces took place on the same limited stretch of inland coastal waters over a number of months showing the limits on the ability of naval forces at that time to project their power inland.

Ossabaw Sound was a large inlet on the Georgia coast, several miles south of where the Savannah River, which flows past Savannah, Georgia, empties into the Atlantic Ocean. The Greater Ogeechee River, which empties into Ossabaw Sound, was navigable for relatively large ships for a number of miles inland. Although major port facilities did not exist on the Greater Ogeechee, the river was close enough to a rail line to make it useful both as a port of refuge for blockade runners and as a terminus where blockade runners could deliver goods for transfer by rail to other parts of the Confederacy. On July 26, 1862, the Federal blockading forces heard a rumor that the blockade runner *Nashville* was up the Greater Ogeechee River with a load of cotton and was waiting for an opportunity to come out. Confederate deserters later confirmed these rumors.

Four Federal gunboats made a reconnaissance up the Greater Ogeechee River on July 29, 1862, and discovered Fort McAllister on the left bank of the river, on high ground that commanded between 1.5 and 2 miles of the river. Fort McAllister was built of earth and sand with a substantial parapet and traverses. The fort contained eight guns arrayed over 400 feet of parapet that faced the river—one 10-inch mortar, one 32-pounder rifle, one 8-inch columbiad, one 42-pounder smoothbore, one 10-inch smoothbore and three 32-pounder smoothbores. The Federal gunboats observed that the river was obstructed with pilings immediately opposite the fort. The fort and the gunboats traded fire for about 90 minutes, but the gunboats ultimately withdrew. The Federals reported that removing the pilings under the "point blank" fire of the fort would have demanded sacrifices that were not warranted. Two Federal gunboats and a mortar schooner made another attack against Fort McAllister on November 19 with a similar result. A third attack, made on January 27, 1863, was different because the ironclad *Montauk* led the Federal forces. The *Montauk* was one of the *Passaic*-class monitors that were built immediately after the fight between the *Monitor* and the *Virginia*. The *Montauk* was slightly larger than the *Monitor*, and she contained a number of improvements that had been suggested by the experience of both building the *Monitor* and her first combat. The *Montauk*'s most significant difference from the *Monitor* was her armament. The *Monitor* had been armed with two 11-inch Dahlgren smoothbores. The *Montauk* was armed with an 11-inch and a 15-inch Dahlgren smoothbore. After the fight between the *Monitor* and the *Virginia*, Assistant Secretary Fox insisted that the new monitors be armed with a larger cannon. Commander Dahlgren, the Navy's ordnance expert, acquiesced in Secretary Fox's instruction and produced a 15-inch smoothbore gun. The *Montauk* was commanded by Commander John L. Worden. As Lieutenant Worden, he had commanded the *Monitor* in her fight against the *Virginia*. In this attack on Fort McAllister, the *Montauk* approached to about 150 yards from the obstructions in the river while the other Federal gunboats remained about 1.25 miles astern. The Confederates had added torpedoes to the obstructions in the river. Federal forces had observed these obstructions in advance of the attack and had staked out a position below the pilings and the torpedoes from which the *Montauk* could fire on the fort. The result of the attack was the same as before: the fort did not suffer any serious damage, and the Federal gunboats withdrew. Shots from the fort struck the *Montauk* a number of times, but she also did not suffer any serious damage. The Confederates did not report any casualties. The *Montauk*'s chief engineer reported that the smoke from her guns was drawn through the boat by the blowers and was discharged through the furnaces without causing much irritation to the crew. The temperature inside the *Montauk* hovered at about 103 degrees during the fight. The chief engineer deemed the performance of the equipment as "satisfactory."

The *Montauk* and the other Federal gunboats renewed their attack on February 1 with somewhat improved results. Fort McAllister reported one dead and several wounded. The parapet in front of one of its guns was demolished, and a couple of the guns were damaged. The damage to the fort was repaired overnight. The *Montauk* suffered some damage from the Confederate fire. Several bolts that held the iron armor in place were broken, and one of the engineers aboard stated that the bolts were made from iron of an inferior quality called "cold short." Being "cold short" meant that the iron was brittle at room temperature as a result of containing too much phosphorous. The crew found that *Montauk*'s turret was difficult to turn until it was raised slightly and then revolved freely. A shell fragment or some other debris may have become wedged between the turret wall and the deck, which may have impeded rotating the turret, and lifting the turret may have freed it. The *Montauk*'s smokestack also suffered a number of hits. On another ship, damage to the smokestack might have impaired her steaming ability, but the *Montauk*'s blower maintained a steady draft and kept the fire in the boiler burning hot. The increased damage suffered by both Fort McAllister and the *Montauk* suggested that the gunners on both sides were becoming more experienced with their guns.

On February 27, 1863, Admiral Du Pont, the commander of the South Atlantic Blockading Squadron, determined to send the monitors *Passaic*, *Patapsco* and *Nahant* to attack Fort McAllister. The *Montauk* was along, but, having already exercised her guns against the fort, she was instructed to remain in reserve. That very evening, an accident happened. Just above Fort McAllister, the Greater Ogeechee River turned 90 degrees to the left and then made a turn of 180 degrees to the right so that the upper portion of the river—known as Seven-Mile Reach—crossed the T of the section of the river that passed just in front of Fort McAllister. The land between these two sections of the river was low and marshy. On the evening of February 27, the *Montauk* caught sight of the steamer *Nashville*, which had run aground at Seven-Mile Reach. The next morning at daylight, the *Montauk* moved up in front of the fort, just below the obstructions, and found the *Nashville* still aground. As the fort fired on the *Montauk* and the other Federal vessels fired upon the fort, the *Montauk* fired 11- and 15-inch shells into the *Nashville* across the intervening marsh at a range of 1,400 yards. The *Montauk* moved up and reduced the range to 1,200 yards. In about 20 minutes the *Nashville* was in flames. A memoir of service aboard the *Montauk* recalled:

> We fire our last shot at three minutes after eight o'clock, having fired fourteen times; and

as the smoke clears away from this last shot, we can see the flames bursting out around her paddle-boxes, issuing in great sheets from the fore-hatch, creeping up the foremast rigging, and gaining aft.

At 35 minutes past 9 o'clock, the *Nashville* blew up.

The explosion was amidships, and the column of flame and smoke, like the discharge of a huge gun, shot up into the air, higher than her trucks, carrying with it the charred and broken timber and burning bales of cotton.... In a few moments another explosion of less extent took place, shattering and opening the stern of the steamer. Her masts, that had stood through it all like black specters, now toppled and came down; the flames gradually lessened; the long black column of smoke wound its way up to a cloud which had grown until it overshadowed the heavens; and nothing remained but the stem and the iron wheels.

A second accident occurred as the *Montauk* withdrew. In her prior attacks on the fort, the *Montauk*'s captain, having been warned about the presence of torpedoes, had kept his boat at least 200 yards below the obstructions in the river. This position kept him clear of torpedoes and still in close range of the fort on the left bank of the river. With the *Nashville* aground at Seven-Mile Reach, his target was more distant, and the obstructions and the torpedoes lay between the *Montauk* and the *Nashville*. A Confederate report of the action states that when the *Montauk* moved forward to close the range, she passed over the torpedoes without detonating any. The report was wrong. As the *Montauk* withdrew from the action, a torpedo exploded under her bottom, nearly amidship, beneath one of the boilers. The explosion broke a cast iron pipe, 4 inches in diameter, that passed from the boiler through the hull. Water started coming in at a rate greater than the pumps could handle. The *Montauk* was steered to a shallow place in the river and permitted to settle on the bottom. This stemmed the flow of water into the boat and allowed the pumps to empty her out. An inspection showed that a permanent repair was not possible with the means at hand, but a temporary repair could be made. When this was done, the *Montauk* returned to Port Royal where permanent repairs were expected to take 10 days to complete. The *Montauk* had been lucky because a larger torpedo would have caused more extensive damage and because the blast took place almost directly under one of her ribs, where the iron was the thickest and the architecture of her construction was best able to withstand the force of the explosion. The experience made the *Montauk*'s people aware that damage to the boilers put them at risk of injury or death by scalding since the vessel's design did not permit an easy means of escape. The risk of injury and death by scalding was shared by all the men who served in the immediate vicinity of the boiler on any steam-powered vessel. The contents of any boiler were under great pressure and breaching the boiler's wall would cause the water within—heated to a temperature above boiling at normal air pressure—to convert into steam all at once and vent explosively. Service aboard the monitors expanded this risk to the entire crew inasmuch as those vessels lacked easy passage above deck.

On March 3, 1863, the monitors *Passaic*, *Patapsco* and *Nahant* launched their joint attack on Fort McAllister. With the *Nashville* destroyed, the principal reason for the attack was to give the crews experience with their novel vessels. Eight hours of firing cut immense holes in the face and the traverses of the fort, but the captain of the *Passaic* believed that the damage done could be repaired with one night's work. In his report, the captain said he did not believe that the fort "can be made untenable by any number of ironclads which the shallow water and narrow channel will permit to be brought into position against it." The captain observed that continuous fire was necessary when attacking earthworks, firing by day to destroy the works and firing by night to prevent repairs. The monitors generally suffered minor damage from the fire returned by Fort McAllister. The exception was the effect of a mortar shell that landed on the deck of the *Passaic*, a place where the armor was thinnest. By chance, the mortar shell struck a beam that supported the deck plating. Had the shell struck elsewhere, and had the shell contained gunpowder rather than sand, it would have caused more extensive damage both to the deck and possibly to the hull.

A Confederate engineer at Fort McAllister also observed that the large Federal guns did not do much damage to the sand fort. He also observed that many of the shells fired by the Federal guns passed over the fort. The fort was set on a rise, and the Federal monitors were firing from water level. They needed to elevate their guns in order to direct them at the fort. A shot placed near the parapet—the top of the fort's wall—would produce noticeable damage to the wall or hit one of the fort's guns. A shot near the bottom of the fort's sand wall likely would bury itself in the sand. By contrast, a shot or shell hitting the bottom of a masonry wall likely would damage the wall and weaken the support for the portions of the wall above. Thus, Federal gunners firing on Fort McAllister were apt to aim their guns at the parapet, and misses and glancing blows would continue sailing up and over the fort. Inasmuch as the fort was on a rise above the water, the Confederate guns exposed themselves to Federal fire only when they were in firing position. The recoil of the Confederate guns' discharge would have pushed those guns back behind the parapet and out of sight of the Federal guns. If the Federals at Fort McAllister had had the experience of Admiral Porter's attacks on Wilmington, North Carolina, in 1864 and 1865, they might have held their fire until the Confederate gunners pushed their guns into firing position. The geography at Fort McAllister permitted only a few gunboats to fire on the fort, while the geography at Fort Fisher permitted many. Thus, at Fort Fisher, many vessels, including the *New Ironsides* with her large broadside, could fire frequently enough to suppress the Confederate fire from Fort Fisher. At Fort McAllister, by contrast, the small number of gunboats could not suppress the Confederate fire, so leaving a monitor's turret pointed toward the fort for an extended period of time invited the Confederate gunners to attempt to do what the Federal gunners might have done: hold their fire to target the enemy's weapons—in this case the open embrasures in the monitor's turret.

 The experience of the monitors attacking Fort McAllister confirmed what Admiral Du Pont and his captains already believed about their ironclad vessels: they possessed exceptionally strong defensive qualities but were deficient in their offensive abilities as compared to other gunboats. The attacking Federal ironclads did not possess the firepower to overwhelm Fort McAllister. As the Confederate engineer observed, enough ironclads could not be brought into the narrow waters before Fort McAllister to overwhelm it. The fort had enough large guns to hold its own against the Federal gunboats and their 15-inch guns. The attack by Federal gunboats, not assisted by land forces, did not threaten to overwhelm the fort or to cut it off from supplies or reinforcements. Early in the war, Confederate troops at Cape Hatteras, North Carolina, and Port Royal, South Carolina, fled when their earthen forts came under fire from large numbers of Federal gunboats. Although large numbers of gunboats might suppress a fort's fire for a time and eventually disable most of its weapons, gunboats alone—even ironclads armed with 15-inch guns—did not possess sufficient force to defeat the fort.

Over much of this same period of time, the Federal Navy Department was sending to the South Atlantic Blockading Squadron most of the coastal monitors and other seagoing ironclads then in commission and was pressuring Admiral Du Pont to use them to sail into Charleston Harbor and demand the surrender of the city. The resulting engagement did not reflect well upon either the admiral or the Navy Department.

3. Exceptional Commercial Circumstances

The blockade running during the Civil War went through several different phases that were distinguished by the types of vessels that were running the blockade and the places from which those vessels came. The initial phase was a continuation of the coastal traffic that had been active before the war. Moving cargoes by water along the coast was less expensive than transportation overland. During 1860 and 1861 approximately 13,200 vessels engaged in the coasting trade; roughly 27.3 percent of the vessels in the coasting trade in 1860 employed steam. The records of the captures (see Chart 3-1) evidence the persistence of this trade, but the captures do not tell us what the overall volume of coastal traffic was at this time. Clearly it dropped off as the blockading squadron grew and the threat of capture increased.

Chart 3-1: Captures of Coasting Vessels
The following records of captures include the date of the report and, where available, the description and name of the vessel as well as its cargo and destination. Atlantic Coast: May 14, 1861, schooners *Mary Willis*, *Delaware*

EXCEPTIONAL COMMERCIAL CIRCUMSTANCES

Farmer and *Emily Ann* from Richmond bound for Baltimore with tobacco; May 26, 1861, American schooner *Iris* bound for Baltimore with naval stores; May 26, 1861, Confederate schooner *Catherine* bound for Baltimore with naval stores; Jun. 8, 1861, small sloops from Smithfield to Norfolk with provisions. **Gulf Coast:** Jun. 22, 1861, sloop *President Filmore* with Confederate coasting license from New Orleans, turned away at Pensacola and bound for Key West; Jun. 25, 1861, schooner *Trois Freres* with Confederate papers from New Orleans bound for Mobile with salt and oats, schooner *Olive Branch* with Confederate papers from Mobile bound for New Orleans with 100 barrels of spirits of turpentine, schooner *Fanny* from New Orleans bound for Mobile with 602 bars of railroad iron, schooner *Basilde* with a Confederate coasting license from New Orleans bound for Mobile with 30,000 bricks; Aug. 28, 1861, schooner *Tom Hicks* from Galveston bound for Port Lavaca with lumber, schooner *General T.J. Chambers* from Calcasieu Bay bound for Galveston with lumber; Aug. 8, 1861, sloop *Charles Henry* of Mobile, previously warned off Pensacola, New Orleans and Mobile; Sep. 25, 1861, schooner *Cecilia* from Sabine River bound for Berwick with 56 pairs of shoes and 14 passengers; Jul. 6, 1862, schooner *Sarah* from Vermillion Bay, Louisiana, bound for Sabine with 73 barrels of molasses and two hogsheads of sugar.

In addition to the coastal traffic, sailing ships continued to visit Confederate ports from overseas. Even though steam-powered vessels regularly carried passengers and cargoes across oceans, most of the world's cargoes continued to be moved by sail. Larger sailing vessels made visits to Confederate ports only in the earliest months of the war, and their masters withdrew them from this trade when Federal steam-powered vessels arrived on blockade stations. Smaller sailing vessels continued to challenge the blockade throughout the war (see Chart 3-2). The number of such captures on the Atlantic Coast fell off in 1863 and 1864, suggesting that this type of blockade running became less attractive as the blockade became stronger. Blockade running by sailing vessels persisted to a greater degree on the Gulf Coast, especially in the areas patrolled by the East Gulf Blockading Squadron, which remained the weakest of the four Federal blockading squadrons throughout the war.

Chart 3-2: Captures of Sailing Vessels
The following records of captures include the date of the report and, where available, the description and name of the vessel as well as its cargo and destination. Atlantic Coast: May 12, 1861, ship *General Parkhill* from Liverpool bound for Charleston; May 14, 1861, ship *Argo* from Richmond bound for Bremen; May 17, 1861, American bark *Star* from Richmond bound for Bremen with tobacco; May 25, 1861, Confederate bark *Sophia* from Rio de Janeiro bound for Richmond; May 30, 1861, Confederate schooner *Lynchburg* from Rio de Janeiro with coffee; Jun. 11, 1861, Confederate brig *Hallie Jackson* from Cuba to Savannah with molasses; Aug. 31, 1861, Confederate ship *Amelia* from Liverpool bound for Charleston with assorted contraband; Jun. 26, 1861, Confederate bark *Sally Magee* from Rio de Janeiro to Richmond; Jul. 11, 1861, Confederate brig *Amy Warwick* from Rio de Janeiro with coffee. North Atlantic Coast: Jul. 20, 1861, Confederate schooner *Velasco* from Cuba with sugar; Aug. 6, 1861, British brigantine *Sarah Starr* from Wilmington bound for Liverpool with naval stores; Aug. 24, 1862, British schooner *Mary Elizabeth* from Nassau bound for Wilmington with salt and fruit; Oct. 11, 1862, British schooner *Revere* from Nassau captured off Wilmington with 800 sacks of salt, 100 barrels of pork and other cargo; Nov. 3, 1862, British schooner *Pathfinder* from Nassau captured off North Carolina with salt, boots, shoes and other cargo; Mar. 24, 1863, British schooner *Mary Jane* from Nassau captured off North Carolina with salt, soap, flour and coffee; May 2, 1863, British schooner *Wanderer* from Nassau bound for Beaufort with salt and herring; Nov. 7, 1863, British schooner *Herald* from Nassau captured off Wilmington with 350 bags of salt and 125 kegs of soda; Jan. 9, 1865, schooner *Triumph* from

Norfolk captured in North Carolina with salt. **South Atlantic Coast:** Nov. 7, 1861, British schooner *Fanny Lee* from Darien, Georgia, bound for Nassau with rice and tobacco; Apr. 3, 1862, schooner *E.J. Waterman* captured near Savannah with coffee; Dec. 13, 1861, schooner *Sarah and Carolina* captured off Florida bound for Nassau with 60 barrels of turpentine; Jan. 7, 1862, British schooner *Prince of Wales* from Nassau destroyed near Georgetown, South Carolina, with salt and oranges, British schooner *Island Belle* from Nassau captured near Charleston with sugar and molasses; May 3, 1862, British schooner *British Empire* from Nassau captured near Matanzas, Florida, with provisions, dry goods, medicines and other cargo; Mar. 31, 1863, British schooner *Antelope* from Nassau captured near Charleston with salt; Apr. 20, 1863, British schooner *Minnie* from Nassau captured near Charleston with 850 sacks of salt, 30 bags of pepper and one cask of coffee; Jan. 3, 1864, British schooner *Sylvanus* from Nassau captured near Doboy Sound, Georgia, with salt, spirits and cordage; Jul. 9, 1864, British schooner *Pocahontas* from Charleston bound for Nassau with 53 bales of cotton and 299 boxes of tobacco; Dec. 3, 1864, British schooner *Mary* from South Carolina bound for Nassau with 77 bales of cotton, 50 boxes of tobacco and one barrel of turpentine; Jan. 28, 1865, and Feb. 3, 1865, Confederate schooner *Coquette* captured at anchor up the Combahee River with 74 bales of cotton. **Gulf Coast:** Jun. 25, 1861, Mexican schooner *Brilliante* bound for New Orleans with 600 barrels of flour and two dismounted guns; Sep. 13, 1861, Mexican schooner *Soledad Cos* from Tampico bound for Galveston with coffee; Oct 17, 1861, British schooner *Edward Barnard* from Mobile bound for Nassau with 600 barrels of spirits of turpentine; Oct. 4, 1861, British schooner *Ezilda* bound for Barataria Bay with contraband, British schooner *Joseph H. Toone* from Havana bound for Barataria Bay with arms and ammunition; Dec. 30, 1861, schooner *Gipsey* captured in Mississippi Sound with cotton; Feb. 3, 1862, schooner *Major Barbour* captured off the coast of Louisiana with gunpowder, niter, sulfur, percussion caps and other cargo. **East Gulf Coast:** Mar. 7, 1862, Confederate schooner *Anna Belle* captured about 250 miles southeast of Pensacola; Sep. 29, 1862, British schooner *Isabel* from St. Marks with cotton; Mar. 13, 1863, British schooner *Surprise* from St. Marks to Havana with 207 bales of cotton; Oct. 30, 1863, British schooner

Meteor from Havana bound for St. Marks with assorted cargo; Feb. 1, 1864, British sloop *Racer* from New Smyrna, Florida, bound for Nassau with 20 bales of cotton; Nov. 16, 1864, Confederate schooner *Badger* from St. Marks bound for Havana with 25 bales of cotton, 9 bales thrown overboard; Feb. 15, 1865, schooner *Delia* captured near Bayport, Florida, with pig lead and sabers. **West Gulf Coast:** Mar. 9, 1862, British schooner *Cora* from Apalachicola bound for Havana with 208 bales of cotton; Jul. 6, 1862, Confederate sloop *Elizabeth* from Havana bound for Sabine with 20 pipes of rum, 20 barrels of coal oil, 10 boxes of gin, two casks of red wine, two barrels of castor oil, three barrels of medicines, two boxes of medicines and 25 pigs of lead; Mar. 25, 1863, schooner *Clara* from Havana bound for Mobile with general cargo; Aug. 22, 1863, Swiss schooner *Wave* from near Galveston with 80 bales of cotton; Feb. 12, 1864, British schooner *Louisa* from Havana bound for Brazos with gunpowder, Enfield rifles, salt, cigars and whiskey; Oct. 19, 1864, British schooner *Annie Virden* from San Bernard, Texas, with 73 bales of cotton; Jan. 14, 1865, Confederate schooner *Josephine* from Galveston bound for Matamoras with 134 bales of cotton; Apr. 22, 1865, British schooner *Chaos* from Galveston with 170 bales of cotton.

An article in the *Nassau Guardian* (reprinted in the *New York Times*) set forth "a complete list of all arrivals [in Nassau] from the Confederate ports since the commencement of the Federal blockade." This information illustrated several changes in blockade running that occurred during the first year of the war: the growth of the volume of blockade running; the increasing use of steamships; the rise of Charleston as a center of blockade running; and the increasing importance of cotton as a cargo through the end of the first year of the war. In more than six months from June 1861 through December 1861, 23 vessels—21 sailing ships and two steamers—arrived in Nassau from the Confederate ports as shown in Table 3-3, carrying cargoes of rice, lumber naval stores and cotton. In just over three months from January 1862 through early April 1862, 35 vessels—25 schooners and 10 steamships—arrived in Nassau from the Confederate ports as shown in Table 3-1 (some vessels made two or more voyages). None of the vessels arrived from Wilmington or Savannah—the Federals occupied the principal water access to Savannah in November 1861. Of the 26 vessels arriving from Charleston, 24 carried cotton, including all nine steamships that sailed from Charleston.

TABLE 3-1: SHIPS ARRIVED IN NASSAU		
Departed	Jun. to Dec. 1861	Jan. to Apr. 1862
Wilmington, NC	4	0
Georgetown, SC	1	3
Charleston, SC	6	26
Beaufort, SC	1	0
Savannah, GA	7	0
Fernandina, FL	1	1
Jacksonville, FL	3	2
St. Johns, FL	0	3
Totals	23	35
Source: *New York Times*, Apr. 28, 1862.		

Most of the vessels that were captured or destroyed attempting to run the blockade were sailing ships as shown in Table 3-2.

TABLE 3-2: SHIPS CAPTURED OR DESTROYED IN THE BLOCKADE			
	Total Number of Ships	Number of Sailing Ships	Percent of Sailing Ships of the Total
Captured	1,149	939	81%
Destroyed	355	260	76%

Notwithstanding the greater number of captures, the persistence of blockade running in sailing ships suggests that it remained sufficiently profitable to justify the risks involved, unlike privateering, which petered out early in the war. During the course of the war, steamers carried the greater shares of the cotton exported from the Confederacy (260,000 bales)—and by extension a comparable portion of the inbound munitions and other supplies—but the amount carried by sailing ships (60,000 bales) was not inconsiderable.

A report submitted shortly after the end of the war by Rawson W. Rawson, British governor of the Bahamas, gives a somewhat different picture of the composition of the blockade running flotilla at least insofar as it was viewed from Nassau. Governor Rawson's report includes statistics that emphasize the importance of steamers as blockade runners and says that while smaller sailing ships were numerous in the early years of the blockade, their numbers dwindled in the latter years, as shown in Table 3-3, which is taken from his report.

TABLE 3-3: VESSELS ARRIVING AND DEPARTING NASSAU				
	Arrived from Southern States		Departed for Southern States	
Years	Steamers	Small Sailing Vessels	Steamers	Small Sailing Vessels
1861	2	2	3	1
1862	32	74	46	109
1863	113	27	173	48
1864	105	6	165	2
1865	35	-	41	-
Total	287	109	428	160

This is inconsistent with the evidence of the captures of sailing vessels noted above, and perhaps the discrepancy reflects faulty record keeping by harbor masters and customs officials in Nassau or the greater ability of smaller vessels in a port busy with the business of blockade running to evade official notice.

Cotton was compressed into bales for shipment. Before the war, the weight of a cotton bale was about 400 pounds, although the size and weight were not uniform, and the average bale in 1860 was closer to 450 pounds. During the war, some of the cotton was compressed further into bales that weighed more than 500 pounds—the average weights of bales reflected in one invoice ranged from 493 pounds to 648 pounds. Compressed bales were described as being about half the size of ordinary bales. Only some cotton presses possessed the power to make compressed bales, so some portion of the cotton was shipped in ordinary bales. Accordingly, any estimate of the volume of cotton shipped based upon the number of bales must be very rough.

Another early phenomenon of blockade running involved sending steam-powered vessels from Europe to the Confederacy and back. The vessels used in this trade were those that had been built and launched before the start of the war, and they consisted generally of two types: steamers with larger carrying capacities that did not emphasize speed, and steamers with lighter drafts and greater speed. The window of opportunity for the larger and slower vessels to run the blockade into Confederate ports closed fairly early. The steamer *Bermuda* (211 feet long and 21 feet in depth) reached Savannah in September 1861, and the steamer *Fingal* (178 feet long and just over 18 feet in depth) arrived there with a cargo of munitions in November

1861. The steamer *Gladiator* (195 feet long and 16 feet in depth) loaded with munitions arrived at Nassau in December 1861, but the presence of a Federal vessel deterred her from continuing to a Confederate port—smaller steamers took her cargo to the Confederacy. Z.C. Pearson and Company, a British firm organized for blockade running, purchased nine steamers, but, between May 4 and August 4, 1862, the Federals captured six of them, and a seventh ran ashore and was destroyed. The losses forced Pearson into bankruptcy in 1862.

The experience of the *Gladiator* set the example for the more enduring phase of blockade running in which large steamers and sailing ships carried cargoes to neutral possessions off the Confederate coast where the cargoes were transferred to smaller, faster steamers that ran the blockade into the Confederate ports. One of the first smaller steamers to run the blockade was the *Theodora*, which was 177 feet long and just over 11 feet in depth. The Bahamas were closest to the Confederacy's Atlantic ports. Bermuda, more distant from those destinations and farther to the north, was less favored. Cuba, then a Spanish possession, was nearest to the Confederacy's Gulf ports. The Confederacy's Atlantic ports were closest to the European markets for cotton. Many blockade runners preferred to use the British possessions as bases for blockade running, but, when circumstances provided an advantage to using the Gulf ports, they also used Cuba as a base.

Dividing the traffic between Europe and the Confederacy into two separate parts with two separate commercial fleets had several advantages. First, the large freighters could carry large cargoes more economically than smaller vessels over the longer leg of the journey. Second, the smaller vessels had a better chance of running the blockade without being captured. Third, sending cargoes from Europe on the neutral vessels to neutral destinations provided some assurance that the vessels and their cargoes would not be seized by Federal blockaders and condemned as a prize by a United States admiralty court. Challenging the blockade directly—by attempting to enter a Confederate port or by exiting a Confederate port—made the vessel and its cargo subject to capture. A neutral vessel at sea was still subject to the belligerent's right of visitation and search, but cargoes sent from a neutral port to a neutral destination were generally exempt from seizure and condemnation. This protection was not absolute. The steamer *Bermuda*, having sailed from Britain to Bermuda, was captured en route to Nassau. Her cargo of about 1,000 tons of contraband included gunpowder, cartridges, artillery and other munitions, and her papers indicated that they were headed to the Confederacy, which made her liable as a prize. The practical effect of this capture was to encourage blockade runners to exercise greater care in composing their bills of lading to conceal the ultimate destination of their cargoes.

Not all the cargoes carried into the Confederacy through the blockade originated in Europe, nor were all the blockade running vessels and their principals Confederates or Europeans. Before the war, interstate trade, including between the northern and southern states, was substantial on rivers, canals, railroads and coastal waters. Such trade persisted with the coming of the war until both the Federal and Confederate governments took steps to cut it off. Some illicit trade between the warring sections persisted but was reduced from what it had been before the war. Through much of the first half of the 19th century, New Orleans served as an entrepot for the produce of the states upriver that did not have an easy access to the port cities on the Atlantic Coast. As railroads changed transportation patterns and provided such access to the eastern states and the Atlantic Coast, New Orleans became primarily a cotton port although the provisions to feed itself continued in large degree to come from upriver. The start of the war cut this source of supply. Thus, when Admiral Farragut captured New Orleans in 1862, he found the city with short supplies of food.

The profits to be made from trading with the enemy were too great a temptation for some. The Federal government, at various times and to varying degrees, authorized the acquisition of Confederate cotton to appease European nations—whose cotton industries were being harmed by the cotton shortage—to provide raw cotton to northern cotton mills to enable the production of clothing and military items, such as tents, and as a form of patronage to the wealthy and powerful interests in northern states to cement their political support. A putative humanitarian justification of some trade was advanced since Confederate communities that the Federals had captured needed to be provisioned to prevent starvation since the war had cut them off from normal sources of supply. In almost all cases it seems, the arrangements provided great profits for those controlling the trade and served as conduits for the flow of provisions and goods through the captured regions into the Confederacy.

Northern goods and provisions also entered the Confederacy through the blockade. Cargoes purchased for shipment to Canada, Bermuda, the West Indies or even Europe ended up as cargoes bound for the Confederacy, sometimes with the knowledge of the seller, and sometimes without, although in the latter case perhaps under circumstances that should have created reasonable suspicions. Northerners were active in blockade running as well. A Confederate merchant active in the blockade running trade spoke of encountering northerners engaged in the trade, professing to be from England or Canada, but he said a moment's conversation betrayed them by their "puritan hypocrisy" and their "nasal twang." In general, he noted, they sold their goods at better prices than the Confederate swindlers, whom he referred to as "Southern yankees."

A later phase of blockade running saw the introduction of steam vessels that were built for the purpose of running the blockade, the construction of each such vessel being prompted by the potential profit in blockade running and the lack of suitable vessels to engage in the trade. These vessels were smaller and shallower than ordinary steam-powered freighters; they had low profiles, and they were generally faster, having powerful engines to drive a limited load. Some had collapsible smokestacks to reduce their profiles further. A number of British shipbuilders were located in Scotland along the banks of the River Clyde, and they provided a number of purpose-built blockade-running vessels. On December 13, 1864, the *Edinburgh Scotsman* reported that "A small fleet of blockade-runners has been passing up and down the Clyde during the last few days." One was a veteran blockade runner that "receiving some damage in the trade, has come home for repairs," and another was a new steamer, a sister ship of another veteran. The *Scotsman* continued:

> Several other new vessels are finishing and building for this trade, and up to the 13th the

total number of vessels, almost all of which are new paddle steamers, that have sailed from the Clyde to run the blockade during the present year, amounted to 60. They average from 400 to 1,200 tons, and were manned with crews of from 20 to 50 men each. The total cost of these vessels would be close upon £500,000.

These vessels possessed characteristics that should have increased the chances of success, but they operated during the latter years of the Civil War in an environment that made blockade running more hazardous: a larger blockading force employing tactics that improved the capture rate. The net result was an offset: the blockaders made some captures, but they were not able to end blockade running so long as a major Confederate port remained open.

The voyage into a blockaded port consisted of three distinct parts—leaving the neutral possession, crossing the open sea and entering the blockaded port—and the outward voyage exposed the vessel to the same kinds of risks but in the opposite order. International law prohibited a belligerent from pursuing or stopping a neutral or enemy vessel within a neutral's territorial waters, which extended 3 miles from the shore, and the presence of foreign naval vessels provided some assurance that the Federals did not impinge upon neutrals' rights or molest Confederate vessels. The Federal desire to curtail blockade running was offset by its wariness of antagonizing foreign governments. Although United States consuls on the islands collected information about blockade runners, they could not communicate it to nearby Federal vessels in a manner timely enough to permit an interception. If Federal vessels were known to be in the vicinity, the blockade runners simply postponed their departures.

Crossing the open sea was the longest part of the voyage. Nassau in the Bahamas was most convenient to the Atlantic ports of the Confederacy—it was about 491 nautical miles distant from Savannah, 503 nautical miles from Charleston and 555 nautical miles from Wilmington. For a steamer making 10 knots, these voyages would have taken more than two days. The distances from Bermuda were slightly greater—about 747 nautical miles to Wilmington, 799 nautical miles to Charleston and 863 nautical miles to Savannah, with the cruising time at 10 knots running a bit more than three days. Havana was convenient to the Confederacy's Gulf ports—it was about 555 nautical miles from Mobile, 621 nautical miles from New Orleans and 770 miles from Galveston. The amount of time required to make any of these voyages depended upon the speed of the vessel, the weather, the encounters with blockaders along the way and the captain's plans for running the blockade at the Confederate port. A blockade runner traveling in either direction spent the daylight hours of two or more days during which they could be seen at a distance by a Federal vessel and pursued.

To an observer at the surface of the ocean, the horizon is about a mile away. To a lookout stationed on a masthead 100 feet above the water, the horizon is about 13 miles away. Sailing ships could be seen about 20 miles—their masts and sails could be seen above the horizon even when their hulls were out of sight. Steam-powered blockade runners took steps to be less conspicuous even in daylight. They kept a low profile whenever possible by removing spars and masts—although they generally retained some mast to keep a lookout aloft—and they were painted a gray color that made them more difficult to be seen at a distance. They burned anthracite or semibituminous Welsh coal whenever it was available since it produced little smoke that could betray their presence at a distance. The smoke from a vessel burning bituminous coal produced a plume of smoke that could be more conspicuous than a sail. One blockade-running captain said that crossing the open sea represented the period of greatest danger, and he opined that the Federals might have been more successful in capturing blockade runners if they had stationed a line of fast steam vessels about 10 to 15 miles apart inside the Gulf Stream.

Another blockade-running captain wrote of sailing his schooner, the *Rob Roy*, from Havana to Galveston. Due to light winds, the voyage took 18 days. When becalmed or in light winds during daylight, the captain kept the sails furled to lessen the chance of being spotted—indeed, several Federal cruisers approached, some as close as 5 miles, but they did not observe his vessel. Later he made the same voyage as the sailing master of the *Phoenix*, a twin-screw steamer with a flat shallow build that carried a larger cargo and made the passage in four days.

If the danger in crossing the open ocean derived from the length of time the blockade runner was at risk of discovery, the danger of entering or leaving a Confederate port grew out of the geography and the assemblage of Federal vessels, but the direction of travel modified the risk. The guns of the Confederate shore defenses generally protected the bar, and the blockading vessels took stations beyond the bar, although sometimes within the range of the Confederate guns, in order to maintain their own freedom of movement. While the blockaders occupied positions outside the bar, a blockade runner exiting a Confederate port in the dark of night could navigate a channel with some caution before he drew close enough to the blockading vessels to risk being seen or heard. Once beyond the bar, the blockade runner generally ceased to be concerned by the depth of the water. If detected and pursued into open water, the blockade runner had a good chance of escaping by losing the pursuer in the darkness. If seen in the morning light, the blockade runner's best hope was superior speed. An inbound blockade runner, if detected, could be chased only a portion of the distance into the harbor before the blockaders came into range of the Confederate shore defenses. On the other hand, if the blockade runner increased his speed upon being detected, he faced a greater risk of running aground. A grounded ship might be destroyed by gunfire or pulled afloat and captured by Federal blockaders; if protected by Confederate guns (if aground near a fort or covered by a mobile battery), some or all of the cargo might be salvaged, and the vessel might be refloated.

The blockade runners took steps to reduce the chances of being discovered. Preferring night and the dark of the Moon, they chose the time and direction from which to pass among the blockading vessels. On nearing a Confederate port in daylight, the blockade runner could slip into an isolated inlet or simply anchor at a distant place and wait until dark. Outbound blockade runners might have information about the positions of the blockaders provided by Confederate lookouts, and they shared information about the number, location and movements of blockading vessels. Some blockade runners ran parallel to the shore in the belief that the surf would mask the noise of their paddlewheels and engines. Some side-wheel steamers installed canvas curtains around their paddlewheels to cut down on the noise.

A sailing ship relied primarily upon stealth to pass the blockade. It did not have any machinery or paddlewheels to betray its presence by sound, and, on a moonless night, especially if the weather was bad, its sail would have been nearly invisible. On one trip to Galveston, the schooner *Rob Roy* was driven dangerously near a shoal; she lowered her sails and used sweeps (long oars) to propel herself to the channel; once there she raised her sails again. On another trip, during bad weather, the *Rob Roy* sent a small boat ahead of her with a kedge anchor, which it dropped, and the crew on board the *Rob Roy* pulled a line to draw the vessel forward to the kedge; the process was repeated until she passed through.

Steamships used stealth when possible, but, if noise and smoke betrayed them, they also had resort to speed. One captain who ran the blockade only in steamships said that he was often fired upon when passing into or out of a Confederate port. The master of the schooner *Rob Roy*, however, made four voyages to Texas and was fired on only once, during an unavoidable daytime entrance. Later, when making his first run through the blockade on a steamship, he commented:

> I must confess that I felt a little nervous at first; the steamer seemed so much longer and more unwieldy to handle in the narrow channel than the little schooner. I soon found, however, that this was more than compensated for by the advantage the steamer had in a steady command of headway, and her readiness to turn in any direction without regard to the direction of the wind.

Although the blockade runner at sea was a solitary enterprise, blockade runners received assistance from a number of sources, not the least from the Federal blockaders. During the first several months of the blockade, in keeping with United States Navy regulations, all the blockaders on station kept a lighted lantern aloft, which both served the blockade runners as navigation lights and permitted them to steer clear of Federal vessels. A Federal consul in the Bahamas reported learning that the blockade runners were using the lights as navigation aids, and, to avoid the blockaders, the Navy Department ordered the lights extinguished. The noise made by the engines of the blockading vessels also betrayed their presence; a donkey pump on the USS *Housatonic* could be heard over a mile away. (A donkey pump or a donkey engine was an auxiliary device that either started a larger engine or performed a function other than propelling the ship.) The blockaders used signal lights and rockets to communicate the presence and direction of a discovered blockade runner, and the blockade runners hoisted their own signal lights and fired off rockets to confuse their would-be pursuers. One blockade-running captain wrote:

> I ordered a lot of rockets from New York. Whenever all hands were called to run through the fleet, an officer was stationed alongside of me on the bridge with the rockets. One or two minutes after our immediate pursuer had sent up his rockets, I would direct ours to be discharged at a right angle to our course. The whole fleet would be misled, for even if the vessel which had discovered us were not deceived, the rest of the fleet would be baffled.

Blockade runners also received assistance from ashore in the form of landmarks and covering fire. The Confederates disabled some peacetime landmarks, such as lighthouses, but others, such as geographic features, remained, and wartime conditions created still others. The Confederates were chronically short on salt, important as a food preservative, and saltworks, whose fires burned day and night, provided a point of reference on the shore. The Confederates had signal lights at Charleston, Wilmington and Mobile. At Wilmington, the signal was placed atop the Mound Battery, a large manmade hill at the southern end of Fort Fisher, an enormous earthen fort built by the Confederates. (The signal is referred to in one blockade runner's memoir as a range light. Pairs of range lights mark different locations and different elevations near the entrance to a harbor. When the vessel observes the lights in a vertical line, the vessel is in the channel.) The coastal forts kept the blockading vessels at some distance from the harbors they protected. The assistance at Wilmington also included a battery of two rifled horse-drawn field guns that went to the site of beached blockade runners and protected them from the Federals.

At times blockade runners shared information, cooperated with one another or were required to share scarce resources. A group of captains who ran the blockade out of Havana agreed that if they encountered one another while at sea they would sail in opposite directions to avoid attracting attention. One imagines as well that blockade runners or Confederate harbor officials provided information about the disposition of Federal blockaders. As blockade running at Wilmington increased, the demand for pilots and others, such as fishermen who knew the local waters, led to a bidding war for their services; local officials stepped in and created a central office that assigned pilots to blockade runners.

From the outset, blockade running was primarily a private commercial venture, and, although the Confederates encouraged it by various activities, initially they let others bring cargoes through the blockade. One early exception was the *Fingal*, which Confederate agents in Britain purchased to carry munitions. The *Fingal* was intended to carry cotton back to Europe, but the increased Federal presence effectively closed the port and trapped her. The Confederates regarded the *Fingal* as too slow and distinctive to challenge the blockade force, and they converted her into an armored warship.

Yet the Confederates also impeded blockade running by allowing diplomatic objectives to override economic necessity. Cargoes shipped from the Confederacy consisted largely of produce items, such as tobacco, rice and naval stores. The most sought-after Confederate produce was cotton, but the Confederates withheld their cotton from the international market through 1861 in an effort to pressure the European powers into recognizing the Confederacy. The cotton embargo was not official government policy, but it was effective nonetheless. Whereas over 1.4 million bales were shipped from the major cotton ports from September 1860 through January 1861, fewer than 10,000 bales were shipped from September 1861 through January 1862. Larger amounts of cotton were shipped from the Confederacy starting in the spring of 1862, which increase coincided with an increase in the number of steamers arriving in Confederate ports.

Even before the end of the embargo, the Confederates looked at cotton as a source of funds for their purchases overseas, but consistent with the posture they had taken with respect to blockade running—that it was an activity to be left to others—they proposed to sell certificates that represented ownership of cotton being held in the Confederacy for delivery at the price it commanded in the Confederacy. The Confederacy possessed most of the world's supply of cotton, but, possessing little manufacturing capacity, it had limited domestic demand for raw cotton. Cotton was selling in Europe at much higher prices, and the differential offered substantial profits that made blockade running attractive. In October 1862 the French banking house Emile Erlanger & Co. proposed to underwrite a public offering of bonds on behalf of the Confederate government. The terms offered were not particularly favorable to the Confederate government—the large discount diminished the proceeds that the Confederates could expect to receive. Nonetheless, the political connections of Erlanger & Co. in Europe motivated hopes of recognition—the same hope that had motivated the cotton embargo—and made the proposal attractive within the Confederate government. A feature that made the bonds an attractive investment was that they could be redeemed at the option of the holders for Confederate cotton in Confederate ports. While negotiations over the terms of the Erlanger loan continued, the Confederates ceased issuing cotton certificates to prevent competition with the Erlanger issue, and Confederate agents in Europe ran low on funds, which curtailed their purchases. The Erlanger bonds were finally issued by March 1863.

By early 1863 the Confederates were becoming dissatisfied with their own hands-off attitude toward blockade running. Many stands of arms had been shipped to the Bahamas, but less than half had reached Confederate ports. Several hundred tons of iron plates, intended to armor the ironclads being built in Charleston, had not been transported from the Bahamas through the blockade. At mid-year, General Pierre G.T. Beauregard, the Confederate commander of the Department of South Carolina, Georgia and Florida, let it be known that any blockade runner that refused to carry a portion of the iron plate would not receive clearance to carry any cargoes out of Charleston. At the end of the summer, the Confederate secretary of war stated that if satisfactory carrying arrangements could not be made with blockade runners, then cargo space would be impressed. Increasing monetary inflation in the Confederacy made holders of produce unwilling to sell for Confederate currency or notes, and commissary and quartermaster agents increasingly impressed goods and forced sales at prices set by the government. Accordingly, the threat of impressment of cargo space was both real and imminent.

Just before the end of 1862, the Confederate Ordnance Bureau purchased four light draft steamers, and in January 1863 they began running between Bermuda and Wilmington, carrying munitions on the inbound journey and cotton on the return. Even with its own fleet of blockade runners, the Ordnance Bureau continued to hire space on private vessels to move its shipments between Bermuda and Wilmington. Through late August 1863, the Ordnance Bureau steamers had completed 22 round trips without a loss. Other bureaus of the Army and the Navy began sending cotton through the blockade to fund overseas purchases. Although they incurred the risk of loss, the proceeds from selling cotton at European prices increased their purchasing power. The Confederate Navy also purchased two blockade runners, but each was lost on its first attempt to run the blockade. The Confederate Navy's third blockade runner, the *Coquette*, had better luck. Although the Navy sold her to private owners, she continued to run the blockade.

Colin J. McRae, the Confederate fiscal agent sent to Europe to supervise the allocation of the Erlanger loan proceeds, advised that the Erlanger loan and the cotton certificates did not yield to the Confederacy the full economic potential of the cotton crop, and he recommended that the sale of such securities cease and all contracts payable in cotton in the Confederacy be annulled. Further, he urged that all cotton be shipped for the account of the Confederate government and shipment for the government's account should travel in government vessels. He went on to propose that a British banking house should purchase or build steamers to run the blockade on behalf of the Confederate government, the vessels to be paid for with the cotton they carried on their first voyages at the local Confederate price; the bargain price of cotton on the first voyage was to be compensation for the financing.

At the time, the Confederate government did not possess enough vessels to take over blockade running entirely, but it entered into contracts with Fraser, Trenholm and Company of Liverpool and John K. Gilliat and Company of London to have 14 vessels built and financed with cotton. It also entered into an agreement with Alexander Collie and Company of London to provide quartermaster, ordnance and medical supplies in exchange for cotton. Collie was to provide four steamers to convey the supplies and cotton. During 1864 and early 1865, Fraser, Trenholm and Company launched six blockade runners, and Collie and Company launched four, that together made 28 successful runs into or out of Confederate ports on 39 attempts. The steamships, all iron-hulled side-wheelers, varied in size from over 1,000 tons and capable of carrying 1,000 bales of cotton to just over 500 tons and capable of carrying 500 bales of cotton. The smaller vessels drew 5 or 6 feet of water, depending upon their load (350 bales versus 500 bales), which made them able to visit smaller harbors. The other eight vessels were not completed in time to take part in blockade running.

The Confederates attempted to exercise control over the blockade-running activities of others with two laws passed in February 1864. The first prohibited the importation of luxuries and items considered not necessary or of common use. Whereas such items had been subject to import duties of up to 25 percent, the new law made all the cargo aboard subject to forfeiture and required the violator to a fine of double the value of the luxury items. The luxury goods were likely to command a greater potential profit per pound, and the prohibition sought to force blockade runners to carry cargoes with greater utility to the Confederacy.

The second law prohibited the export of "cotton, tobacco, military and naval stores, sugar, molasses and rice" except pursuant to government regulations, and these required the vessel to allocate half her tonnage to the government; it also required the shipper to give a bond that the vessel would return to a Confederate port "with reasonable dispatch" with half the cargo tonnage allocated to the government. The regulations provided that the government's cotton or tobacco bore freight at five pence sterling per pound (about $0.21 in 1864 US dollars or $2.31 in 2019 US dollars), payable on delivery in coin or sterling exchange, and return freight was at £25 per ton (about $250 in 1864 US dollars or $2,774 in 2019 US dollars) payable on delivery in cotton at 10 pence per pound. In addition, the regulations required a bond that one half the net proceeds of the outward voyage either would be spent in goods to be shipped to the Confederacy within 60 days of unloading or paid to the Confederacy; amounts paid would be reimbursed by delivery to the shipper of cotton at the rate of 10 pence sterling per pound at a Confederate port. The regulations made blockade running a less attractive proposition, but, for those who continued in the business, they created substantial economic incentives to remain engaged in it. Efforts by the blockade runners to modify the regulations failed, but most of the blockade runners remained in the business.

By the summer of 1864 the three principal Confederate ports in operation were Wilmington, Charleston and Mobile. The Federals captured Mobile Bay in August 1864, and they captured Fort Fisher, which closed the port of Wilmington, in January 1865. Charleston fell when General Sherman's Army, having marched from Atlanta to the sea at Savannah, arrived at Charleston in February. The war was not yet over, but the *New York Times* observed:

> All the British blockade-running ports, from Nova Scotia to New-Providence, are filled with enterprising craft, laden with Confederate ammunition and stores, but unable even to venture within sight of the coast-line. . . . Not less pointed . . . are the latest official returns of the British Board of Trade. For the last-reported month there was a decline of exports of £2,258,963 sterling, as compared with the reports for the same month of the previous year.

4. Planning the Blockade and Blockading Tactics

The forces opposing the blockade—the blockade runners, of which there were Confederates and others; the Confederate government; and foreign nations, particularly Britain and France—had separate agendas that both overlapped and conflicted. Where their agendas overlapped, the forces cooperated, but, where the agendas conflicted, the forces pulled in separate directions. The efforts to run the blockade were essentially atomized—each vessel made a solo effort to carry its cargo to or from a blockaded port, even if the vessel was a part of a larger corporate undertaking that operated several vessels, such as a shipping company or the Confederate ordnance department.

By contrast, a single force supported the blockade—the Federal government spearheaded by the Navy—and mounting the blockade was a collective undertaking that required constant coordination of its parts if the effort was to be at all effective. The magnitude of the undertaking required a vast increase in the size of the Navy in terms of the number of vessels it operated; the number of men and officers required for those vessels; the coordination of the activities of those vessels and men; the logistical support to keep those vessels and men on station; and the establishment of bases close to the blockade stations to provide support for vessels and men. And, as it turned out, the Navy's services were required in other aspects of the war, some of which related to the blockade, such as assisting in the capture of ports, and some of which did not.

Although President Lincoln proclaimed the blockade two days after the surrender of Fort Sumter, the imposition of the blockade proceeded slowly because of the variety of crises and tasks that demanded the Navy's attention and the small size of the fleet. The crises and tasks included:

- A flotilla sent to supply the garrison at Fort Sumter at Charleston arrived to find the Confederate bombardment in progress. After the fort's surrender, the vessels carried the garrison north.
- In a similar standoff at Fort Pickens at Pensacola, Federal warships anchored nearby helped keep Confederate forces at bay.
- The Navy cooperated with the Federal Army in holding the forts at Key West and the Tortugas.
- After the commander of the Federal Army posts in Texas surrendered to Confederate forces, a Federal warship escorted transports to remove loyal Federal men and officers from the Texas coast.
- As Virginia moved toward secession, the Federal Navy Department sent a force to destroy the facilities, vessels

and munitions at the Norfolk Navy Yard. The attempt had limited success; the Confederates obtained many cannon, and they salvaged the USS *Merrimack* for conversion into the ironclad CSS *Virginia*.
- When a riot in Baltimore cut rail service to Washington City, Navy vessels escorted transports carrying troops to protect the capital.
- The Navy also patrolled the Potomac River to prevent the Confederates from obstructing the channel and cutting off water access to Washington City.

At the start of President Lincoln's term, the Federal Navy consisted of 42 vessels that were in commission and on active service. The 12 vessels of the Home Squadron were stationed in northern ports, Pensacola or Vera Cruz. The other 30 vessels were on foreign station or on the Pacific Coast. An additional 18 vessels were not in service and awaiting repair, not counting those lost at Norfolk. All but one of these were back in service by mid-July 1861. Most of the vessels on foreign station were summoned home and had arrived by mid-November 1861.

The president's proclamations in April 1861 made public the intent to mount a blockade. The blockade of each port was accomplished by stationing at least one vessel before it and giving formal notice in the port. The Federals established the blockade of Charleston, Mobile, New Orleans and Savannah during May and Wilmington on July 21. The two outlets of the Cape Fear River giving access to Wilmington were separated by an island and shoals that made the cruising distance between them over 40 miles. The sole Federal vessel stationed at Wilmington divided her time between the outlets in an attempt to demonstrate that both were covered.

Initially, Secretary of the Navy Gideon Welles directed individual vessels to their specific duties, including the blockade, but, during May 1861, he took steps to create a command-and-supply structure needed to support a much larger force. He formed the Atlantic Blockading Squadron and the Gulf Blockading Squadron and assigned to their commanding officers the responsibilities of directing the actions of the squadrons and obtaining the desired results. He also directed the chiefs of the Navy's administrative departments to determine how to supply the blockading squadrons. The chiefs recommended that supply ships from northern ports should supply each squadron at its rendezvous—a sheltered port preferably near its area of operations.

In June 1861 Secretary Welles assembled a committee (a "conference" in the parlance of the day) to outline the operations to be undertaken by the Federal blockaders. The participants—two Navy officers, one Army officer and the superintendent of the United States Coast Survey—brought considerable professional experience and information to the task, although with some inherent qualifications. The Navy members' professional experience included service during the Mexican-American War, but novel technologies that the Civil War would introduce, such as armor and exploding underwater mines, were beyond the compass of their prior experience. The Army brought information about the construction and armament of the preexisting coastal defenses in the southern states, but this information became outdated as the Confederates supplemented those defenses. The superintendent of the United States Coast Survey brought geographical information about the Confederate coast and adjacent waters—by 1861 over 3,700 miles of coastline had been surveyed and over 45,400 square miles of coastal waters had been sounded—including the bars before the harbors and the channels that ran through them. The United States government had developed this information as an aid to navigation, which meant that all but the most recent was generally available. By contrast, once the war began, the Coastal Survey printed a series of volumes titled *Notes on the Coast* that dealt with the coast from Delaware Bay to Texas. These supplemented the commercially published sailing directions with discussions of items of military importance and each included 8 to 12 charts with sailing directions and information about winds and tides. To preserve secrecy, the text of these booklets was written in longhand on lithograph stones and printed in the United States Coast Survey offices rather than typeset and printed at the Government Printing Office or a commercial printer.

From July through September 1861, the conference produced a series of reports on the organization and support of the blockading squadrons and the various means of closing Confederate ports. The conference recommended that the Atlantic Coast be divided into two separate commands for efficient administration—the smaller size enabled a commander to visit all his forces over the course of one or two days. (The sailing distance from the mouth of the Chesapeake Bay to the border between North and South Carolina—the area patrolled by the North Atlantic Blockading Squadron—is about 350 nautical miles, and the distance from that state border to Cape Canaveral—the area patrolled by the South Atlantic Blockading Squadron—is about 356 nautical miles. A ship cruising at 10 knots for 48 hours could easily cover either distance.) To maintain the blockade, the Federals needed harbors on the Confederate coast to serve as supply depots, repair facilities, and places of refuge during bad weather. The northern portion of the Confederate Atlantic Coast was proximate to Federal ports that could serve the purpose. The southern portion of the Atlantic Coast and Gulf Coast needed local depots. For the southern Atlantic Coast, the conference examined Fernandina, Florida, and three other sites that were located closer to Charleston and Savannah. They considered both the advantages offered by each site as a prospective naval station, as well as the effort required to capture and hold it. Fernandina was located to the south of the principal Atlantic Coast cotton ports, the area likely to see the most intensive blockading. (Florida was largely undeveloped with few rail connections below the panhandle, so the lower two-thirds of the peninsula was unlikely to see much blockade running.) Although symmetry spoke in favor of Fernandina—Secretary Welles may have suggested it, which may explain why it figured prominently in the conference's report—proximity to major Confederate ports spoke in favor of the other sites.

The conference's final reports focused upon the Gulf Coast with most of the attention given to New Orleans. Although the Mississippi River provided the primary water route to New Orleans, other waterways provided access as well so that controlling the river did not close access to the port. The conference did not recommend attacking New Orleans, but it considered the defenses on the river south of the city (Forts Jackson and St. Philip) and the benefits of occupying the Head of Passes—the point where the river divided into several different streams that fanned out into the Gulf of Mexico. The conference recommended seizing Ship Island, located off the northern coast of the Gulf about midway between New Orleans and Mobile, the two major Gulf Coast ports, as a depot and harbor of refuge. (At the time, the Confederates held Pensacola, although the Federals retained possession of Fort Pickens. The Federals occupied Pensacola when the Confederates abandoned it in May 1862.) The conference considered Galveston the only point on the Texas coast that was worthy of attention in the blockade.

The conference also recommended different methods of controlling the coast based upon the nature of the local terrain and the likelihood that an area would attract a substantial amount of blockade running. It recommended obstructing various inlets with hulks and seizing other inlets to prevent their usage. Bays and inlets that were too large to be obstructed would be "blockaded . . . in the usual manner" or captured to prevent their becoming active ports, although, as it turned out, blockading in the "usual manner" would require some readjustment to then-current conditions. Active major ports required that groups of blockading vessels be stationed there permanently; lesser ports also required constant attention but by fewer vessels; the lightly inhabited and uninhabited stretches of the coast—such as Florida south of Fernandina—could be patrolled by cruisers in constant motion.

It did not address the issue of traffic running between Mexico and the Confederacy, perhaps because the blockade could not be mounted there. Inasmuch as Mexico was a neutral nation, international law prohibited the Federals from imposing a blockade. Although the law did permit Federal cruisers to stop private vessels in international waters and inspect their cargoes for contraband being shipped to or from the Confederates, they were prohibited from interfering with cargoes nominally shipped by neutrals to neutrals, and they were prohibited from stopping neutral or Confederate vessels in the territorial waters of a neutral nation. The Federals would have the least success in interfering with this trade.

Even as the conference completed its work, the Federal Navy Department started implementing its recommendations. The Army agreed to a joint operation to secure a harbor on the Confederate Atlantic Coast, and the Navy Department selected Captain Samuel F. Du Pont, the president of the conference, to command on behalf of the Navy. The Navy Department also divided the Atlantic Blockading Squadron into two separate commands, and it appointed Captain Du Pont to command the southern squadron, elevating him to the grade of flag officer. To preserve secrecy, the Navy Department instructed Flag Officer Du Pont to select the target, and, on November 7, 1861, Federal warships captured the Port Royal Sound in South Carolina.

With only a small fleet at the start of the war, the Federal Navy Department seized upon the conference's suggestion to close Confederate harbors with obstructions. The department purchased 45 old sailing ships, loaded them with stone—referred to as the "stone fleet" or the "stone blockade"—and sent them to be sunk in the harbor entrances of Charleston and Savannah. The *New York Times,* writing under the headline "Doomed Cities," predicted that the effect would be permanent.

Confederate General Robert E. Lee, then overseeing southern coastal defenses, wrote, "This achievement, so unworthy any nation, is the abortive expression of the malice and revenge of a people which it wishes to perpetuate by rendering more memorable a day hateful in their calendar." The British lodged a formal protest with the United States government, and the British Foreign Secretary was reported to have stated in the House of Lords:

> The permanent destruction of a harbor was not an act of war or man against man, or of nation against nation, but it was an act of war against the bounty of Providence, which vouchsafed harbors for the advantage of commerce and for the civilizing influence of intercourse between one people and another.

The international outrage was not diminished by the recollection that the British had obstructed Savannah during the Revolutionary War or by the assurance of Secretary of State William H. Seward that the obstructions were temporary and would be removed upon the restoration of the Union.

The turmoil raised over the stone fleet was apparently well founded. Professor Alexander D. Bache, superintendent of the United States Coast Survey and a member of the blockading conference, wrote to Captain Du Pont, "I think well of [Assistant Secretary of the Navy Gustavus V.] Fox's idea of closing up that entrance [to Charleston Harbor], and will bring you the evidence to-night." Professor Bache could speak with as much authority about the conditions on the ocean floor around the United States as anyone in the world. On December

17, 1861, the Federals scuttled 16 hulks in the Main Ship Channel at Charleston. The hulks were arranged "so nearly to overlie each other that it would be difficult to draw a line through them in the direction of the channel which would not be intercepted by some one of the vessels." The hulks began to sink in the mud almost immediately, and nothing could be seen of them at the end of a week.

Just as the neutral powers feared that the stone fleet would be too effective, some Federal naval officers feared that it would not be effective at all and, therefore, should not be attempted. Flag Officer Louis M. Goldsborough, in command of the Atlantic Blockading Squadron in 1861, carried on a correspondence—lasting from mid-October through mid-November—with Lieutenant Reed Werden who was charged with scuttling stone-laden hulks in the inlets of North Carolina. Lieutenant Werden attempted to convince Flag Officer Goldsborough that sinking the stone fleet would be useless, and Flag Officer Goldsborough, with diminishing patience, instructed the junior officer that he had his orders and should carry them out. What is curious about this extended conversation is that Lieutenant Werden was not a "volunteer" officer—a civilian mariner brought in to swell the Navy officer corps for the war—who might not have been acclimated to the military chain of command. He was, rather, a career officer who had joined the Navy in the 1830s and eventually became an admiral. Flag Officer Goldsborough had his orders from Secretary Welles. Even if he believed that the stone fleet would not work, a chance remained that it might work and thus advance the cause of the war and the mission of his squadron. If it did not work, the attempt would be a waste of resources, but those resources already had been expended in assembling the stone fleet and taking it to the southern coast, and the effort would not otherwise hinder the Federal war effort or the mission of his squadron. Flag Officer Goldsborough stood to lose more by arguing common sense than he did by simply following his orders. Flag Officer Goldsborough, the senior officer present, understood the politics of the situation even if his subordinate did not. A hurricane moving up the North Carolina coast in November 1861 destroyed a number of the stone vessels that were intended for the North Carolina sounds. About a week later the Navy sank two stone vessels in Ocracoke Inlet. In transmitting the correspondence and the reports of the sinkings to the Navy Department, Flag Officer Goldsborough added the endorsement, "I beg to transmit these letters to the Department. They refer to the blocking up of Ocracoke Inlet, which work, I am happy to say, has at length been accomplished."

Ideas of how the blockade could be mounted and maintained changed radically during the war. Secretary Welles initially believed that an effective blockade of a principal port could be imposed by one large steamer and several smaller gunboats. Although this statement might have reflected some wishful thinking based upon the small number of vessels then available, it also reflected a modestly updated version of how blockades were conducted in the age of sail, with a few large vessels moving up and down a distance from the blockaded port. This belief was shared by the officers of the blockading conference—who wrote that major ports would be "blockaded ... in the usual manner"—and the officers who commanded the blockading squadrons during the earlier part of the war. Flag Officer Silas H. Stringham, allocated 15 vessels to blockade the entire Atlantic Coast, wrote that an additional 12 to 15 efficient and seagoing vessels would be needed to make the blockade of the Atlantic Coast "perfect and strict." In exactly the same way, Flag Officer William W. Mervine, allocated 23 vessels to blockade the entire Gulf Coast, wrote that a total of 31 steamers of various sizes and four sailing vessels were needed to make "an efficient blockade of the coast of the Gulf."

On November 26, 1861, Captain James L. Lardner, the senior officer off Charleston, had three steamships and a sailing ship stationed off the main entrance to Charleston and two adjacent inlets, and he reported, "I believe these places to be effectually blockaded," although he acknowledged that his ships would be obliged to go to sea in the event of an easterly gale. "In that time it is possible for a vessel to get to sea. I can hear of none likely to attempt it." Several days later, Flag Officer Du Pont reported that the blockade of Charleston was so rigorous that the fishermen had been driven in, denying the city its supply of fish. He continued:

> There is one obstruction to a constantly efficient blockade that can neither be removed nor over come, and that is fog.
>
> The vessels that lie in wait to run the blockade, having skillful pilots, and being desperate in their attempts, cannot but sometimes succeed under the favor of fog or darkness.

This sanguine appraisal was dashed by consular reports from the West Indies and Cuba in November and December 1861 and newspaper reports of large numbers of vessels arriving there from Charleston and Savannah. The United States State Department also received information that lanterns hanging in the blockaders' mastheads—the running lights required by Navy regulations—betrayed their positions and provided navigational aid to the blockade runners. The lights were extinguished. Prompted by these reports, in late March 1862, the Navy Department instructed Flag Officer Du Pont to convene a court of inquiry to investigate the circumstances of any evasion of the blockade, and, in early April, the United States Senate instructed the Committee on Naval Affairs "to enquire whether there has been any laxity on the part of our naval officers charged with the blockade" of the south Atlantic Coast and Charleston.

The experience of the South Atlantic Blockading Squadron eventually confirmed the consular and newspaper reports—the squadron captured a number of schooners that were spotted during daylight a distance from Charleston; it made a few captures in the immediate vicinity of Charleston; and it reported a few incidents of having seen vessels escaping into or out of Charleston. The several nighttime captures appear to have been accidental—on April 19, 1862, for example, the USS *G.W. Bunt* en route to Georgetown, South Carolina, "fell in with" the schooner *Wave*, carrying 39 bales of cotton, headed from Charleston to Nassau.

Flag Officer Du Pont wrote an upbeat response to the Senate's inquiry, saying:

> I am fully prepared to capture every vessel attempting to run the blockade, so far as my force and circumstances will admit.... With rare exception none but very small craft and two or

> three rebel steamers . . . and mostly under the protection of the night or dense fogs, have been successful.

He stated that only two large steamers had run the blockade, and the rest had transferred their cargoes to smaller vessels. He saw "abundant evidence of the stringency of the blockade in the great scarcity of even the necessities of life, and the high price demanded for both food and clothing." He pointed to his own experience in blockading during the war against Mexico, and he asserted that "no blockade in the history of the world has ever been more effective." With respect to a report that armed Confederate vessels had gone to and from Charleston without being molested, Flag Officer Du Pont wrote that it was "one of those absurd partisan statements of which this rebellion has been so fruitful."

Flag Officer Du Pont was more candid in responding to the Navy Department's order that reassigned the USS *Susquehanna* and its commander, Captain James L. Lardner, at the time the senior officer off Charleston:

> I require more vessels everywhere. The Department is sending me more and more stringent directions in reference to the blockade, directing courts of enquiry to be held for any infractions of it, and the Senate is passing resolutions reflecting on myself and my officers, and I appeal to the justice of the Department if this is a moment to reduce my force and take from me my most efficient ships and most experienced officers.

The combination of darkness and steam gave the advantage to the blockade runners. On a moonless night, the visibility was several hundred yards at best. Flag Officer Du Pont observed in a personal letter, that in Port Royal "we have some twenty vessels anchored around us in the visual proximity, and if it were not for their lights we would not see half of them." Steam power liberated the vessel from the wind and tide, which gave the blockade runner more flexibility in choosing where and when he would attempt to run the blockade.

With Captain Lardner reassigned, Commander John B. Marchand became the senior officer off Charleston. At that time, 10 blockading vessels—nine steamers and a sailing ship—formed a line approximately 26 miles in length, with a slightly greater concentration of vessels before the Charleston bar—5 of 10 blockading vessels were arranged off the Charleston bar in an arc that bowed out away from the bar with an additional sentinel close to the Lawford Channel (the western extension of the Main Ship Channel). Although the vessels nearest to Charleston were closer together, the average distance from one blockader to the next was over 2 miles.

Commander Marchand rearranged the blockading force—six steamers and four sailing ships—to cover a shorter portion of the coast and shifted their positions so that seven vessels formed two concentric arcs with an additional sentinel close to the Lawford Channel. The distance from one steamer to the next was half a mile. One vessel sent a picket boat closer inshore at night. A blockade runner sailing to or from Charleston passed by the picket boat and through two lines of blockading vessels. At dawn two of the steamers stationed close to the shore sailed several miles seaward looking for vessels having come out that night, and another patrol sailed at dusk looking for vessels preparing to enter that night.

The success of the blockaders off Charleston changed about two weeks after Commander Marchand took command. At 4:50 on the morning of May 24, 1862, the USS *Bienville* sighted a steamship about 15 miles off Charleston and heading east. The Moon was a waning crescent with only 14 percent of the visible surface illuminated; the new Moon would occur on May 28. The sighting occurred about a half hour before dawn but about the start of civil twilight (and, therefore, about a half hour after the start of nautical twilight), when artificial light is not needed to carry on outdoor activities. The *Bienville* gave chase and captured the vessel about 6:25 AM. She was the *Stettin*, having sailed from Nassau with a cargo that included saltpeter, pig lead, quinine, tea, coffee and tin plate. She had been stripped of her spars, sails and upper masts, and she was painted a lead color.

At five on the morning of May 26, the USS *Huron,* which was anchored near the entrance to one of the channels through the bar, saw a steamship trying to enter the channel. Upon being spotted, the steamship turned and fled. The *Huron* slipped anchor, gave chase and made the capture shortly after nine. She was the *Cambria*, out of Liverpool by way of Nassau with a cargo of rifles, saltpeter, muslins, medicines and hardware.

At 4:45 on the morning of May 27, the USS *Bienville* was off Bull's Island, about 18 miles distant from the Charleston bar, when she saw a steamship. The *Bienville* gave chase and captured her about an hour later. She was the *Patras* from Britain by way of St. Thomas and Havana, and she carried a cargo of gunpowder, arms, quinine, coffee and other merchandise.

All three captures resulted from a combination of the blockade runners' mistakes and the blockaders being in position to take advantage of them. The blockade runners had sailed when the phase of the Moon was most favorable, but they remained too close to Charleston as the sky started to brighten in advance of dawn. The *Cambria,* at the bar, failed to find the channel and turned to run back to sea. The USS *Huron* was on hand to see her in the improving light. The USS *Bienville* spent most of the night on blockade near the entrance to the harbor but sailed east before dawn in the hope of spotting a blockade runner as the sky grew brighter. At that hour she should have expected to see an outbound vessel with a cargo of cotton rather than an inbound vessel with a cargo of munitions and other goods. As dawn approached, an inbound vessel should have sailed away from Charleston and made another attempt on the next night. All three of the captured steamships were running the blockade for the first time, so the possibility exists that the captains, although experienced masters, did not have experience in running the blockade.

Although the Federal reports of the captures seem to state time approximately rather than exactly, and although different conventions for marking time were used, on the one hand, by the blockaders and, on the other, by the 21st-century United States Naval Observatory (the source of the historical information about the local time of civil twilight and dawn), the proximity of the first sightings to local dawn supports the surmise that the skies were brightening.

The events as the next new Moon approached demonstrated that the changed tactics improved the Federal ability to detect blockade runners. At 10:30 PM on June 19, 1862, the USS *Keystone State* discovered a schooner slipping out of Charleston. The Moon (in its last quarter with 44 percent of its visible disc illuminated) had not yet risen. The schooner slipped out in a squall, which was generally favorable to the blockade runner, but may have permitted her to steer too close to the *Keystone State*. The Federals pursued and about 2 AM captured the *Sarah* carrying 156 bales of cotton to Nassau. At daylight on June 20, the USS *Alabama* captured the schooner *Catalina* that also had come out during the squall carrying 33 bales of cotton. At 2:30 on the morning of June 23, blockaders discovered a steamship heading into Charleston, which slipped past them before they could react. The Moon, which was a waning crescent with 11 percent of its visible face illuminated, had just risen or was just about to rise. Shortly after, another steamer tried to run the blockade. The Federals observed her, and, when the USS *Seneca* fired upon her, she turned and ran to sea. The USS *Keystone State* and the USS *Flag* pursued. The *Flag*, being slower, gave up the chase after about 60 miles. The *Keystone State* gave pursuit over 300 miles but lost the steamer the next night in bad weather.

In September 1862 the North Atlantic Blockading Squadron made similar changes in tactics in their blockade of Wilmington. The three Federal steamers on station off New Inlet were described as anchored in the same positions, night and day, 5 miles off the inlet, 2 miles apart. Acting Rear Admiral Samuel P. Lee, newly in command of the squadron, ordered the blockading vessels to move at twilight and approach as near to the bar as was safe and practicable. He also instructed them to use a steamer as a picket vessel, which created the same layered arrangement used at Charleston, although with fewer vessels. The blockade running into Wilmington remained light at the time, and the blockading force remained small—the distance between the inlets required the squadron to patrol both. The effectiveness of the blockade at Wilmington did not improve until later when the importance of Wilmington rose as a center of blockade running and the number of blockading vessels increased.

Changing the arrangement and movement of the blockading vessels and increasing the observational capacity of the blockading force by the use of picket boats established a pattern of maintaining the blockade that persisted for the remainder of the war. The changed tactics improved the chances that the blockaders would detect blockade runners, which was a prerequisite for making captures. Although the change made the blockaders more capable, darkness and surprise continued to favor the blockade runners, and a large portion of their attempts were successful. The blockading squadron continued to adjust its tactics and to elaborate upon them in the hope of increasing the number of captures and overcoming newly arisen difficulties.

The number of vessels available to blockaders, and steamers in particular, increased throughout the war, which facilitated variations in tactics. At Charleston, Commander Marchand used three sailing ships at anchor as the outermost of the two lines of blockaders. They were his slowest and most vulnerable vessels, and he placed them farthest from the shore batteries. Admiral David D. Porter, who took command of the blockade at Wilmington in late 1864, specified that the slower steamers would compose the inshore blockading line and directed that they would not pursue blockade runners but instead would fire upon them. By late 1864 Admiral Porter had enough steamers to use them in both the inner and the outer lines of blockaders.

The increased resources and revised tactics created new problems. For example, extinguishing all lights at night and moving the blockading vessels closer together increased the incidence of blockaders firing upon one another. The USS *Susquehanna* was fired upon by another Federal vessel on the evening she arrived off Charleston. Two nights later the USS *Susquehanna* fired upon the USS *James Adger* even though the two vessels remained anchored exactly where they had been before sundown. One presumes that the cases of friendly fire were more frequent when vessels got underway in the dark because the commander at Charleston ordered that "when a vessel is underway at night she will hoist three perpendicular lights at her gaff, each at least 8 feet apart; the upper and lower to be white and the middle red." (At the time "perpendicular" had a primary meaning of "at right angles to the plane of the horizon.") The signal was changed two weeks later apparently to prevent fraudulent use by blockade runners. By late August 1862 the instructions called for a different combination of two lights for each day of the week.

The need to confirm friendly identity was balanced against the desire not to give assistance to blockade runners. Running lights were to be hoisted only by vessels when near or approaching the blockading vessels but kept aloft "no longer than absolutely necessary." Blockaders at Charleston also were instructed to recognize one another at night by a flash of light—produced by uncovering and covering a lantern directed so that it could be seen only by the other vessel. The signals changed over time and were used to communicate more information, such as the direction in which a blockade runner was headed. Other blockading squadrons developed night signals for the same purposes.

Although blockade runners ventured from a blockaded port or planned to enter it during darkness, their journeys to or from Bermuda, the Bahamas or Cuba always embraced a period of daylight in the open sea when they could be seen at a greater distance and pursued. When vessels could be spared from the close-in blockade, positioning outside vessels produced additional captures. This tactic produced captures around Charleston and Wilmington. While Commander Enoch G. Parrott was in temporary command of the Charleston blockade, he sent a vessel from the inner blockade seaward to a position just out of sight of the blockaders. He reported, "This has resulted in the capture of the [blockade runners] *General C. C. Pinckney* and the *Mary Teresa,* and I think it would be well to keep a vessel permanently in that position." The blockaders off Mobile also used it but to a lesser extent than at Charleston or Wilmington.

As the blockaders developed the tactics that improved the effectiveness of the blockade, they increased their presence on the water at night by using cutters and launches—oared boats carried aboard their larger vessels—as picket boats. A picket boat could not effect a capture, but it could detect and signal the presence of a blockade runner so that the steamers could react and pursue. The duty was hazardous—the passing blockade runners could run down a picket boat or swamp it, and wind, tides and weather could put the picket in danger of being captured. For defense, the picket boats carried small arms, and some of them carried a boat howitzer. Admiral Lee maintained a strong prejudice against using oared boats as pickets, preferring steamers instead, but his North Atlantic Blockading Squadron still made some use of them. The squadron's USS *Matthew*

Vassar, a sailing vessel, guarded an inlet near Wilmington and stationed a picket boat close to the inlet each night. On several occasions, the picket boat deterred a steamer from entering the inlet, although the lack of wind usually prevented the *Matthew Vassar* from making a capture. Notwithstanding the success of the *Matthew Vassar*'s picket boat in deterring steamers, Admiral Lee believed that the use of picket boats "is a temptation to a dangerous relaxation of vigilance on board the [blockading] steamers." He ordered, "Use picket steamers hereafter."

The various elements of the improved blockading discussed above reflect learning from information provided or experiments that yielded positive results. In late December 1862 Lieutenant Stephen P. Quackenbush was placed in command of the USS *Unadilla* and sent to Charleston, his first posting at an active blockade-running port. On January 27, 1863, at two in the morning, a nearby vessel made a signal that a blockade runner was seen heading into Charleston. Lieutenant Commander Quackenbush prepared to get underway, and about 5 or 10 minutes later he slipped anchor and headed in the direction indicated by the signal, but he did not "get full sight of" the blockade runner and believed she got safely into Charleston. Two nights later, with the prior experience apparently in mind, at 3:15 in the morning the *Unadilla* saw a signal, slipped anchor and got underway immediately. She sighted, pursued and captured the *Princess Royal,* a British steamer with a cargo that included rifled guns, arms, ammunition and two marine steam engines. That not only denied Confederates the benefit of the cargo—the engines were apparently intended for a Confederate ironclad—but also because the Federal Navy purchased both the *Princess Royal* and the two steam engines.

The former became the USS *Princess Royal*, and the latter powered the USS *Kansas*—both vessels saw service as blockaders and made several captures.

The blockaders had at least two tools at their disposal that they did not think to use as we have come to expect. As noted, the Federals used calcium lights during their operations in Charleston Harbor against Fort Wagner, and, at least for a time, they mounted such lights on the monitors stationed in the Main Ship Channel, but they did not use them to detect and track blockade runners until near the end of the war.

The Federals were painfully aware that the Confederates deployed submerged torpedoes as weapons against them, but they do not appear to have considered using them as weapons to close a Confederate port to blockade running. The stone fleet had shown that physical obstructions were not effective to stop traffic. The Federals controlled the coastal waters beyond the range of the Confederate shore guns, so they could have maintained torpedoes as an ultimate deterrent to blockade running. Recalling the international fury that arose in response to the stone fleet, however, one can imagine the reaction if one or more "neutral" blockade runners had been sunk by an underwater mine.

Although the tactics that improved the conventional naval blockade were developed at Charleston, the concentration of forces in and around the city caused the manner in which the blockade was conducted there to become significantly different from the manner in which it was conducted elsewhere.

By April 1863 the Federal Navy Department had sent nine ironclads to the South Atlantic Blockading Squadron—all but two of the ironclads then in Federal service on coastal waters. Admiral Du Pont found the monitors strong in defense but weak in projecting offensive power, and, being fearful of losing any of them to the Confederates, he did not find any use for them in blockading. Secretary Welles prodded Admiral Du Pont relentlessly to use the ironclads offensively. Ultimately, and against his better judgment, the admiral led the ironclads into Charleston Harbor where they fired upon Fort Sumter for about 40 minutes, were fired upon by Confederate batteries, and then withdrew.

In July 1863 the Navy Department replaced Admiral Du Pont with Rear Admiral John A. Dahlgren who cooperated with the Army in operations against the Confederate defenses around Charleston Harbor—monitors and other warships providing support for the ground forces attacking Forts Wagner and Sumter. In addition, Admiral Dahlgren deterred blockade running by pushing the monitors as far inside the harbor as safety and weather conditions permitted—the monitors' armor allowed them to occupy positions under the Confederate guns that would have been untenable for conventional wooden warships, and their low profile made them difficult targets to hit, especially at night. The monitors occupied the Main Ship Channel after dark. A tug brought about six armed launches to serve as pickets to a position above the monitors to protect them from Confederate attacks. On the night of July 19, 1863, a large blockade runner attempting to enter Charleston was forced aground by the monitor *Catskill*, which was anchored abreast of Fort Wagner on picket duty. In his report, Admiral Dahlgren crowed, "The chances of success to such enterprises are materially lessened by our occupation of the main Ship Channel."

In December 1863 Admiral Dahlgren organized a picket force to form an inner blockade consisting of four monitors, two tugs and about eight launches and boats. Two of the monitors were to be on duty each night, one well advanced up the harbor to observe Fort Sumter and Fort Moultrie, and the other farther back but in a position to render support. Two tugs served as tenders to the monitors and stayed on patrol with them. Two of the boats had boat howitzers, and the others had small arms. Half of these boats were on picket duty with the monitors to enforce the blockade and to check the movement of Confederate torpedo boats and picket boats. The other half was a reserve that provided relief during the night. Additional Federal scout boats operated independently of the pickets on the order of the admiral and reported to him directly. In January 1864 Admiral Dahlgren ordered the monitors to supplement their defense against threatened attacks from torpedo boats and submarines (the Confederate submarine *H. L. Hunley* attacked and sank the USS *Housatonic* in Charleston's outer harbor in February 1864) by installing outriggers to suspend rope netting all the length on their sides and by positioning on their decks two boat howitzers on field carriages loaded with canister. A trial was made to place a calcium light on the deck of a monitor, but the commanding officer thought it to be a disadvantage, possibly because the light made the monitor a more conspicuous target in its advanced position.

The combination of the presence of so many vessels inside the bar and within the outer harbor, and the continuance of active operations, curtailed blockade running at Charleston. In both June and July 1863, five steam-powered blockade runners arrived in Charleston. One blockade runner arrived in August and none in September. The Confederates acknowledged that the port of Charleston had been "closed." The Federal inability to break the defenses at Charleston led to a cessation of active operations and a deployment of forces to other places. Blockade running resumed either as a result of relaxed Federal vigilance or increased determination on the part of the blockade runners: two blockade runners arrived in March 1864, none in April and two more in each of May and June. Admiral Dahlgren issued extensive and detailed instructions about the posting of vessels on the inner and outer blockades, about the command responsibilities of the various officers present and signaling information about the movements of blockade runners and the orders for pursuit. Blockade running continued, although at a relatively low level, until the Confederates abandoned Charleston in February 1865 as General William T. Sherman's Army, having marched from Atlanta to the sea, continued its march north through South Carolina.

5. Mounting and Maintaining the Blockade

The experience at Charleston showed that a naval blockade in the steam era could be made somewhat effective—but not absolute—given the technologies that existed at the time. The preparations that made possible the formulation of more effective blockading tactics were the actions taken by the Federal Navy Department. These included (1) increasing the size of the fleet; (2) building ironclads; (3) obtaining qualified officers to command and men to operate the vessels; (4) keeping the squadrons supplied; (5) providing the means to keep the vessels in repair; (6) providing administrative support for the squadron commanders; and (7) deploying the vessels to the theaters where they would do the most good. In four years of war the Federal Navy grew from 42 vessels in commission to over 670 vessels. The ability of the Federal Navy Department to fulfill these tasks suffered from time to time under the challenge of that growth. Equally, the commanders of the blockading squadrons understood that the chance for success of their missions—and their professional advancement—improved with larger forces under their command, and they called for more ships, faster ships and more men.

1. **Increasing the Size of the Fleet.** As the number of vessels increased, the overall character of the fleet changed as well. The prewar fleet consisted largely of deep draft ocean-going ships that could not readily navigate the relatively shallow southern ports. The different types of services that the Navy was called upon to perform required different types of vessels—a fast and lightly armed vessel for blockading duty and a more heavily built and heavily armed vessel for attacking harbor fortifications. As the threat of Confederate ironclads became more palpable, the Federal blockaders, unable at first to get their own ironclads, asked for larger warships. With the exception of the ironclad USS *New Ironsides,* the Federals did not build any steam frigates during the war. The squadron commanders also called for tugboats, torpedo boats and vessels to guard against attacks from Confederate torpedo boats.

At the time of President Lincoln's inauguration, the fleet consisted of 42 vessels in commission. By July 1861 the Atlantic Blockading Squadron had 22 vessels, and the Gulf Blockading Squadron had 21. By early December 1861, Secretary Welles reported that 136 vessels had been purchased and 52 vessels had been built or were under construction for Navy service, and, when these were complete, he expected to have a fleet of 264 vessels. When the Navy took vessels built for civilian purposes into military service, it generally modified them to make them strong enough to support the weight and the strain of firing the guns that would arm them. It also installed a magazine and additional pumps.

The size of the fleet at different times during the war is summarized in Table 5-1. By December 1864 the Federal Navy of 671 vessels included 559 steam vessels (of which 71 were ironclads) and 112 sailing vessels (16 percent of the fleet)—although Secretary Welles emphasized the importance of steamers, the Federal Navy continued to purchase and use substantial numbers of sailing vessels. The speed of the blockaders was challenged by the newer, purpose-built vessels that were brought into service as blockade runners. Eventually, some of these newer blockade runners were captured and brought into Federal service. For example, on August 24, 1864, the USS *Gettysburg*, which previously had been the blockade runner *Margaret and Jessie*, captured the blockade runner *Lilian* out of Wilmington with a cargo of cotton. The USS *Lilian* became a blockader off Wilmington.

TABLE 5-1: SIZE OF FLEET	
Date	Number of Vessels
March 1861	42
December 1861	264*
December 1862	427
December 1863	588
December 1864	671

* Includes vessels under construction.

The rapid expansion of the Navy's fleet generated some controversy. Secretary Welles entrusted much of the ship purchasing to George D. Morgan, a businessman whose judgment he trusted in these matters. Over the course of four-and-a-half months, Mr. Morgan purchased 98 vessels for the Navy Department for a total of about $3.5 million (about $100 million in 2019 dollars) or $900,000 less than the sellers' asking prices. For these services, Mr. Morgan received the standard commission of 2.5 percent from the sellers, which totaled nearly

$90,000 (over $2.5 million in 2019 dollars). Mr. Morgan was the cousin of the incumbent Republican governor of New York, Edward D. Morgan, and he was Secretary Welles' brother-in-law. Both the newspapers and some members of Congress attacked the integrity of the transactions, but Secretary Welles insisted upon Mr. Morgan's integrity and the appropriateness of his commissions. President Lincoln stood beside his Navy secretary, a Senate motion to censure Secretary Welles failed, and the matter eventually died down.

2. Developing Ironclads. Although ironclads were a part of the larger fleet that the Federals built during the war, they were novel and untested weapons, with characteristics that were different from traditional wooden warships and that posed novel challenges and opportunities in their use. These factors justify an extended consideration of this class of weapons.

As a part of the attempt to destroy the Norfolk Navy Yard when the Federals abandoned it on April 20, 1861, they set fire to the USS *Merrimack*, a steam-powered screw frigate built in 1855, and then scuttled her. About a month later, the Confederates raised the *Merrimack* and began to convert her into the ironclad CSS *Virginia*. At roughly the same time as the Federal Navy bureau chiefs advised Secretary Welles on how to supply the blockading squadrons, they also made recommendations about shipbuilding policy and the need for a "mail clad floating battery."

The Federal Navy's ship constructors and engineers said they were too busy with present matters to handle the relatively unknown business of designing and building an ironclad vessel. In early August 1861 the Federal Navy Department published an advertisement requesting proposals "for the construction of one or more iron-clad steam vessels of war, either of iron or of iron and wood combined, for sea or river service of not less than ten nor over sixteen feet draught of water." The Navy board formed to review the proposals—referred to as the Ironclad Board—stated, "Our immediate demands seem to require, first, so far as practicable, vessels invulnerable to shot, of light draught of water, to penetrate our shoal harbors, rivers and bayous." From 17 proposals received, the board accepted three that became the *New Ironsides,* the *Galena* and the *Monitor.* The *New Ironsides* and the *Galena* were conventional wooden warships that were covered in iron. The *Monitor* defied contemporary expectations of how a warship should look. The *Monitor* consisted of a turret—a cylinder large enough to hold two large cannon—that sat on a low, flat raft. The *Monitor* lacked masts, rigging and sails, which were common to warships of the day, even if they used steam propulsion.

The battery of a conventional warship in 1861 generally consisted of many broadside guns and one or more pivot guns fore and aft. With the firepower strongest on the beam, bringing it to bear depended upon the direction in which the ship was headed. Although each broadside gun could be aimed to some extent, the ability of each broadside gun to bear upon a target depended upon the direction in which the ship was pointed. The armament of a conventional wooden warship had its greatest power when it fired on broadside or beam, and it was weakest when it was attacked on the bow or the stern. By contrast, the *Monitor*'s two cannon were mounted in a turret that could turn to fire in almost any direction, regardless of where the bow pointed, making her firepower equally strong—or weak—in all directions. Whereas a conventional wooden warship relied upon firepower for both her offense and defense, the *Monitor* coupled relatively small striking power with substantial shot resistance.

The *Monitor*'s cannon could not fire directly over the bow. The vessel's design placed the pilothouse mostly under the deck in front of the turret. A small iron structure rose above the deck on top of the pilothouse, and the commander or helmsman viewed the waters ahead of the vessel through a small slit in the structure. The *Monitor*'s cannon fired just above the deck, and their barrels did not reach the pilothouse, so firing over the bow subjected the pilothouse and its occupants to the effects of the muzzle blast. Later versions of the *Monitor* relocated the pilothouse to the top of the turret, which did not limit aiming the guns and gave the commander or helmsman a view on all sides of the vessel. Placing the pilothouse below the deck of the *Monitor* was simpler mechanically than placing it atop the turret, and the decision to do so was driven by the desire to finish construction quickly to meet the emergency posed by the *Virginia*'s construction and imminent launch.

A conventional warship provided a substantial target, with its wooden hull rising high above the water and its masts and rigging rising high above the deck—the *Merrimack*, for example, was 300 feet in length overall, and her deck was over a dozen feet above the water. The *Monitor* was 172 feet long and her deck rose only 18 inches above the water; her turret was 9 feet tall and 21 feet across. The turret was the most heavily armored portion of the vessel, consisting of overlapping iron plates built up to 8 inches thick. Other exposed surfaces were covered by iron plate—4.5 inches on the sides and 2 inches on the deck. Thus, the *Monitor* offered a relatively small target, and she was most heavily armored where she was most likely to be hit.

On March 8, 1862, the CSS *Virginia* and several smaller wooden vessels steamed into Hampton Roads in Chesapeake Bay to confront the Federal flotilla on station there. Several of the Federal vessels ran aground and were unable to flee or bring more of their guns to bear. The *Virginia* rammed the USS *Cumberland* and pierced her wooden hull with a massive cast iron beak. The *Cumberland* sank in minutes. The *Virginia* and other Confederate vessels then turned on the USS *Congress* and shelled her, setting her on fire with hot shot. As the tide waned, the Confederate vessels returned to the Norfolk Navy Yard. The *Congress* suffered 120 killed and missing and 26 wounded. The *Cumberland* suffered 121 killed and missing and at least 16 wounded. Four other Federal vessels suffered six killed and 36 wounded.

The news from Hampton Roads sent the senior members of the Lincoln government into a near panic. Secretary Welles' diary recounts that Secretary of War Edwin M. Stanton "made some sneering inquiry" about the *Monitor*, "and when I mentioned she had two guns, his mingled look of incredulity and contempt cannot be described."

The *Monitor* arrived at Hampton Roads that evening and stationed herself in front of the USS *Minnesota*, which was also aground but unharmed. The next morning, when the *Virginia* returned, the *Monitor* went out to meet her, and the two ships fired at each other for over four hours until a shell struck the *Monitor*'s pilothouse, blinding her captain. The *Monitor* was steered into shoal water, and another officer was summoned to take command. By the time that the *Monitor* returned to resume the fight, the *Virginia* was steaming away.

The *Virginia* had sustained some damage in her initial attack on the *Cumberland* and the *Congress*. Neither the *Virginia* nor the *Monitor* substantially damaged the other. The *Virginia* attempted to ram the *Monitor*, but the *Monitor*, faster and nimbler, avoided being rammed. The *Virginia* was armed with shells and heated shot, effective against wooden vessels but less so against an ironclad. (The British succeeded in breaching a 4.5-inch iron plate, and thicker plates as well, by firing a solid metal bolt from a rifled cannon. They also developed exploding shells to be used against armored vessels. The heat generated when the shell hit its target detonated the bursting charge. At first, the shells burst almost immediately upon striking the armor, which diminished the shell's destructive power. Placing the bursting charge in a flannel bag delayed the detonation until the shell punched through the armor and into the interior of the ship. While both the Federals and the Confederates had steel or hardened iron projectiles for use against ironclads, neither possessed an armor-piercing shell.) The *Monitor* was armed with a pair of 11-inch smoothbore cannon, the largest used on the Navy's vessels at the time, but none of the *Monitor*'s shots did substantial damage to the *Virginia*. As noted above, the *Monitor*'s guns used the standard powder charge approved by the Navy regulations but did not use the wrought iron cannonballs specially prepared by John Ericsson. The *Monitor* suffered one wounded (her captain, Lieutenant John L. Worden) and three injured by concussion. The CSS *Virginia* suffered two killed and 19 wounded (including her commander, Flag Officer Franklin Buchanan), all occurring on the first day of the battle and none during the fight against the *Monitor*.

Had the *Monitor* not arrived when she did, the *Virginia* would have captured or destroyed the other wooden vessel that had run aground around Hampton Roads. She would have claimed dominance of the lower Chesapeake Bay, but her influence beyond that was open to speculation. The *Virginia*'s draft was too deep for her to go up the Potomac River and shell Washington, DC, although she might have threatened Baltimore. While her commander considered her unseaworthy, and her range was limited because she was buoyant when loaded with just several days' supply of coal and water, she might have been towed to New York. In open water, the wooden ships of the Federal Navy would have been better able to maneuver and avoid her attacks. Indeed, while armor made the ironclad stronger in defense, it did not make her invincible, as later experience showed. When Federal warships captured or destroyed Confederate ironclads, larger caliber guns appear to have inflicted most of the damage, but the Federal Navy did not have these until later in the war—in March 1862 the *Virginia* might have been a tougher nut to crack.

Before the war ended, the Federals had built or were building 50 additional Federal ironclad vessels on the *Monitor* design, with one or more turrets, for service on the coast or ocean. Nineteen vessels were roughly the same size as the *Monitor* and were intended to patrol coastal areas—these were the *Passaic*-class and *Canonicus*-class monitors. Eleven vessels were substantially larger than the *Monitor* and intended to be seagoing.

Twenty other vessels were smaller than the *Monitor* and intended for shallower coastal and inland waters—these *Casco*-class or light draft monitors were all unseaworthy due to defects in planning and oversight. John Ericsson had provided a simple design concept for the light draft monitors, but Alban C. Stimers, the Navy's chief engineer who oversaw the construction plans and supervised the project, included elaborations and additional performance-enhancing mechanical devices that overwhelmed the original simple concept. These changes increased the weight of the vessels, but Mr. Stimers did not take into account its impact upon their buoyancy. When the hulls were launched without their turrets and stores, their decks were only 3 inches above the surface of the water.

The Federals also built numerous vessels for service inland on the Mississippi River and other western waters. These included iron-plated riverboats, casemented ironclads and monitor-style vessels with one, two or three turrets.

The chronic difficulties faced by the *Passaic*-class and *Canonicus*-class monitors were due in part to their relatively small size. As noted, at the earliest stage of the planning process, the Ironclad Board recognized that a shallow draft would be of primary importance given where the boats were expected to operate. Successful design involves making compromises and accommodations. Armor, steam propulsion and large caliber cannon increased the weight of a warship, which decreased buoyancy, and excess or reserve buoyancy was what kept a vessel afloat. As a result of design compromises, striking power, defensive power and mobility were all diminished, and the reserve buoyancy was small.

To compensate for the limited reserve buoyancy, the monitors had powerful pumps to remove water and drainage channels or gutters called "limbers" in their bottoms that allowed the water to flow to the pumps. The limbers had to be clear, and the vessel had to be properly trimmed, for water to flow through the limbers to the pumps in the stern. The pumps ran on steam, so if the power failed, the pumps failed. One *Passaic*-class monitor, the USS *Weehawken,* sank while at anchor during a moderate gale. The anchor chain passed through the hull in a small hole near the waterline toward the bow of the ship, but a stopper in the hole that was intended to seal out the water had not been put in place. Water entered the hull around the anchor chain near the bow, causing the ship to tip its bow deeper so all the water in the vessel flowed toward the bow, away from the pumps. The crew did not have any apprehension of the danger until 10 or 15 minutes before signaling for assistance, and the vessel sank barely 5 minutes later.

As with any weapon, the monitors' characteristics made them suited for certain tasks and unsuited to others. They were lethal to any conventional wooden warships and could go toe to toe with almost any other ironclad of the time, as the fight between the *Monitor* and the *Virginia* demonstrated. Admiral Du Pont sent his monitors to attack earthen and masonry forts with little effect, and he found little other use for them. Admiral Dahlgren, having replaced Admiral Du Pont, used the monitors to give close-in support for the siege operations that captured Fort Wagner, an earthen fort that guarded an approach to Fort Sumter. Later he also used monitors at night to occupy the channel into the inner harbor, effectively closing the port of Charleston to blockade running.

3. Obtaining Officers and Men. As the sectional crisis developed, and through the early days of the Civil War, a number of United States Navy officers who felt allegiance to the southern states resigned their commissions. Secretary Welles worried that his officers included undisclosed traitors and lukewarm

patriots, and he warned his squadron commanders to be vigilant for treasonous sentiments in the officers under their command and for inefficiency and apathy in the execution of their duties. Secretary Welles' initial problem in the selection of senior commanders was to find men who were both loyal and capable, and he was guided initially both by reputation and seniority. As time passed and he gained exposure to the individual officers, he disregarded seniority and selected younger men who displayed leadership qualities and aggression; he instructed his senior commanders to seek these qualities in the officers in their squadrons; and he also removed from important commands the less vigorous or successful commanders. The Federal Navy grew so large and so fast that the number of career naval officers was not sufficient, and the Federals had to recruit "acting" or "volunteer" officers from civilian life. The 7,500 volunteer officers eventually constituted about 85 percent of the Federal Navy officer corps, although the most important commands usually went to career officers.

Moreover, as the war progressed, the average age of the serving career officers became lower. When Admiral Samuel F. Du Pont led the ironclads in his squadron in a brief attack on Fort Sumter in April 1863, the officers commanding his seven monitors included three captains and four commanders—they had begun their careers in the Navy as midshipmen in the years 1827 to 1839, so by 1863 they had on average about 30 years of professional experience. Six months later in October 1863, the command of the monitors had turned over completely—the officer commanding the eight monitors included two commanders and six lieutenant commanders. Like their predecessors, all these men were career officers, but they became midshipmen between the years 1836 and 1853, and they had on average just over 20 years of professional experience. When Secretary Welles asked about the youth of the lieutenant commanders, Admiral Dahlgren (who had replaced Admiral Du Pont) responded that two commanders had declined to take a monitor when offered; he noted that the squadron had been shorthanded since he arrived, and, since the vacancies occurred "when the duty was most severe," he filled them with the officers present.

Along with ships and officers, the Navy needed men. In 1860 the number of enlisted men in the Navy was about 8,800 (exclusive of Marines), and during 1861 the number rose to about 26,800. By July 1863 the Federal Navy had 34,000 men in service. The Federal draft, and the recruitment laws that set manpower quotas for the states, did not give the states credit for men who joined the Navy, and the laws also did not authorize transfers between the services. In December 1863 Secretary Welles reported to Congress that "the difficulty in procuring a sufficient number of seamen for the increase of the service has been great, and at times vessels have been detained and unable to proceed to sea for want of crews." In March 1864 some 30 to 40 Federal warships were ready for active service but were tied up for lack of crews. Toward the end of the war, bounties of up to $1,000 per man were paid. The Navy had about 51,500 men in service at the end of the war.

Manpower shortages prompted the Navy Department to measures unusual for the times. Even before the Civil War, the United States Navy, unlike the Army, accepted Black men into its ranks, although it confined them to the lower ranks, so it was tapping a larger manpower pool from the outset. Early in the war, Secretary Welles also decided to enlist contrabands.

When Secretary Welles read in the newspaper in June 1863 that the Federal government was feeding from 7,000 to 8,000 Black people on the Louisiana coast, he wrote to Admiral Farragut advising him to enlist the able-bodied men into the Navy and noting that the department would like to send 200 men to the Pacific. Secretary Welles authorized the commander of the West Gulf Blockading Squadron to retain in service enlisted men whose terms were up and directed that no discharges would be issued except upon order of the department.

4. Providing Supplies. The adequacy of supply was itself the primary concern, but supply arrangements focused upon operational and fiscal efficiency as well. Different supply arrangements were tried at different times, and arrangements that kept the blockading vessels away from their stations for the least amount of time were preferred. Initially, blockading vessels had to go to northern ports or Key West for fuel and supplies. The capture of Port Royal provided a base on the Atlantic Coast of the Confederacy, and the Confederates' abandonment of Pensacola and Ship Island provided bases on the Gulf Coast. The Federal Navy Department sent supplies in stores ships that became floating warehouses. The department also sent supplies in blockading vessels returning from northern ports, in the Navy's transport vessels and in commercial vessels chartered by the Navy or operated by contractors who were hired by the Navy to provide supplies to or trade with the blockading squadrons.

Supply was difficult from the first due to the large variety of items that were consumed aboard naval vessels. These included paymasters' stores, including items such as provisions, clothing, candles, stationery, medical supplies, tobacco, soap, beeswax, needles, razors and shaving soap; coal to stoke the boilers of the steamers; engineers' stores, such as lubricating oils, tools and replacement parts for the machinery; ships' stores, such as sails, rigging, chains, flags, paint, lanterns, illuminating oils, anchors and other replacement parts; and ordnance stores, including gunpowder, ammunition and accoutrements for cannon of various sizes and small arms, cartridge bags to hold gunpowder charges, primers to discharge cannon, fuses to ignite shells, related equipment and Coston signal flares.

In addition to the supplies provided by the Navy, officers and men could purchase additional mess items, such as flour and hams from independent vendors, called sutlers, who occupied space in the Navy's supply vessels where they were subject to Navy oversight. Regulations allotted a sutler 1,600 cubic feet of storage space and one stateroom aboard a Navy supply steamer. As the number of blockading vessels grew, the storage allocated to the sutler on occasions was not sufficient to meet the demands of all the vessels visited. Some sutlers operated their own vessels, and, although the larger number of men in the Army made them a more attractive clientele, the sutlers sold to men on Navy vessels as well. Although sutlers were licensed by the government, they sometimes succumbed to the opportunity to profit from illicit trade either by smuggling alcohol to Federal forces or trading with the Confederates. In one instance, the Navy seized a sutler's schooner that had on board "428 dozen cans of strong drink resembling eggnog, marked on the manifest as 'milk,' and on the cans as 'milk drink.'" A schooner arriving at Washington, North Carolina, with a cargo of sutler's stores included on its manifest "snuff, wine, dry goods and hoop skirts." Other vessels, including government vessels, participated in this illicit trade.

Squadron commanders took the departure of a blockader for a northern port as an opportunity to improve the supply of stores and address any manpower shortages for the vessels that remained on station. During 1862 Flag Officer Farragut ordered vessels departing the West Gulf Blockading Squadron to leave behind the ordnance and other stores that they did not need for their voyages. In October 1864 Admiral Porter issued general orders that a vessel leaving the North Atlantic Blockading squadron should "transfer all the men that could be spared, stores, clothing, and provisions to vessels that stand in need of them."

The ration provided by the Navy consisted of items that were capable of being stored at room temperatures for long periods of time by a method such as salting, pickling, smoking, spicing or drying. The sailing bark, the USS *Fernandina*, was stationed along a section of coast that Confederates had abandoned, and her crew supplemented their rations by hunting, gathering and trading with Black people living in the area. The ship's doctor wrote often about food in his diary, perhaps because the variety of the menu was a primary source of relief from the boredom of the duty. On the long, slow journey from the blockade to a northern port, away from a supply of fresh food, the fare aboard the *Fernandina* consisted solely of stored provisions: salt junk (salted beef and salted pork), hardtack and beans, and the doctor complained, "Potatoes and onions are played out. Our diet is a regular prison one."

By 1860 using ice to preserve fresh foods had become common in certain businesses and in middle- and upper-class urban households, and the business of harvesting ice from northern ponds for use in the summer had become substantial. Mechanical refrigeration was in commercial use in Britain, but it had not taken hold in the United States, perhaps due to the abundance and relatively low cost of ice.

Early in the war, the Federal Navy Department purchased two steamers—the USS *Connecticut* and the USS *Rhode Island*—that it sent to deliver fresh beef, vegetables and ice to the blockaders. At first, the Navy Department sent these supply steamers to visit every vessel on both the Atlantic and Gulf Coasts with instructions to deliver a two-day supply of beef and vegetables, and ice to preserve them, on the outward voyage and a comparable amount on the return voyage. As the number of vessels on blockade increased, the number of supply steamers increased as well, and the Navy Department modified their orders, initially to visit the vessels on the Atlantic Coast or the Gulf Coast and, at a later time, to visit the vessels of one of the four blockading squadrons. By January 1865 the Federal Navy had at least five steamers delivering supplies to the blockading squadrons. Although the steamers' principal duty was supply, they were armed, and they pursued and captured blockade runners they encountered during their voyages.

The arrangements for preserving the beef underwent some adjustment to improve efficiency. For its first voyage, the *Connecticut* had 400 quarters of beef hung on hooks in a chill room that was cooled by ice in a room built above it. The captain of the *Connecticut* also built two smaller icehouses in which he stacked alternating layers of ice and beef until the structures were full. After leaving New York, the *Connecticut* did not begin dispensing beef until it reached Charleston, at which point the beef in the chill room remained fit, but about 25 percent of the ice above it had melted. By the time the *Connecticut* reached Pensacola, portions of the beef in the chill room were showing signs of taint. Although the chill room retained enough wholesome beef to distribute to the vessels visited on the outward voyage, a large quantity spoiled and was thrown away. When the *Connecticut* reached Galveston, all the beef remaining in the chill room had spoiled. In the smaller icehouses, the beef and ice remained a frozen solid mass, and the amount of ice had shrunk by only a few inches. The *Connecticut* distributed beef from the icehouses to all the blockading vessels on its return voyage. The *Connecticut* sailed from New York with about 59,000 pounds of beef and 125 tons of ice. The captain estimated that 35,000 to 45,000 pounds of beef stowed in the icehouse method would have been "sufficient for all demands." Not all the lessons of experience were learned well or passed along: when the supply steamer USS *Bermuda* was sent on her first voyage to the Gulf Blockading Squadrons in May 1863, due to faulty preparation of the icehouse, most of the 175 tons of ice melted by the time she reached Key West, and most of the beef she carried was spoiled.

Water was a supply concern as well. Historically, the water supply was a limiting factor on ocean voyages. The use of a condenser to produce potable water removed that limitation, but substituted fuel as a crucial commodity. The *Connecticut* initially lacked a condenser and, apparently having on board only enough for her own needs, was not able to provide water to the sailing sloop USS *Jamestown*. The *Connecticut*'s report noted that the *Jamestown* "was obliged to abandon [her station] and go in quest of water."

In addition to supplies of fresh beef and vegetables, the supply vessels carried other stores and the mail. When the *Connecticut* left New York in January 1862, she carried 60,000 letters and newspapers and 2,500 packages for the personnel of the Federal Navy and the Army. Outbound vessels carried officers and drafts of men to the blockading stations, and homebound vessels carried prisoners, invalids and men whose enlistments had expired. One vessel carried a draft of 650 men. The needs of the blockaders sometimes taxed the capacity of the vessels. In January 1862 the captain of the *Connecticut* was advised that his ship would carry a large amount of gunpowder from New York to the Gulf Blockading Squadron. In his letter to the commander of the Navy Yard, the captain acknowledged the need, but he reported that the ship's magazine was already full. For proper safety, a suitable structure would need to be built to house the gunpowder, but, since his ship also was "crowded to capacity," the captain did not know where that might be. The captain concluded his letter to the commander of the Navy Yard, "I have, in obedience to your order, received the onions; as the only place I can stow them is on the after part of the deck, it is more than probable they will be lost."

With the majority of blockading vessels driven by steam, the coal supply was crucial. Federal forces retained control of Key West at the start of the Civil War, and, by the start of July 1861, the Navy Department had purchased three ships, averaging 1,200 tons each, that were loaded with coal and sent to Key West to supply the vessels of the Gulf Blockading Squadron. When the Federal Navy captured Port Royal in November 1861, it did not build facilities on shore but rather tended to the needs of the South Atlantic Blockading Squadron with coal vessels and other supply and service vessels. Immediately upon the fall of Fort Macon at Beaufort, North Carolina, in May 1862, the Navy established Beaufort as a base for coal and other supplies to support the North Atlantic Blockading Squadron—vessels in need of supplies and repair were ordered to Beaufort and were not to go farther north unless absolutely necessary.

Civilian appointees in principal northern ports, known as "Navy agents," handled certain business and fiscal matters on behalf of the Navy Department, including the purchase of coal. Early in the war, squadron commanders wrote to the Navy agents in Philadelphia and New York to request shipments of coal, but shipments were sporadic—generally too little, which impaired operations, and occasionally too much, which incurred demurrage charges as the vessels that brought the coal were detained for want of a place to offload and store it. In early March 1862 Flag Officer Farragut informed the department that Ship Island was without coal, and the squadron had been forced to borrow 800 tons from the Army. Later that month, Flag Officer Du Pont wrote the department that his steamers needed 3,500 tons to make up their present deficit and were expected to consume coal at the rate of 1,100 tons a week. Although the department gave instructions to address these needs, by September 1862 Flag Officer Du Pont wrote that he was short of coal and would have been without any had he not been able to draw upon the Army's supply. The situation improved somewhat after the Navy appointed an officer to supervise coal supplies in November 1862, although the improvement could have been due in part to the market adjusting to the rising level of demand for coal and for vessels to ship it. As the squadrons continued to grow through the war, their need for coal increased as well. Refueling at the bases was an easier task because it could be done at a pier or in slack water. Refueling on station, although more difficult, meant minimal interruption from blockading duties. Rough weather, especially in winter, could make coaling on station all but impossible. Coal was transferred from ship to ship in baskets or bags.

Blockading tactics affected the demand for coal, and the availability of coal affected tactics. Blockaders kept their steam up to be prepared for an immediate pursuit or to react quickly in the case of a Confederate attack, but this level of preparedness increased the consumption. If fuel was in short supply, the blockading vessels might be forced to spend more of their time at anchor with their fires low—although this conserved fuel, it increased the time that it took for a blockading vessel to get underway. The dependence of the Federal squadrons on coal also made them subject to the fortunes of war waged on land. When the Confederates invaded Maryland in 1862 (which culminated in the battle of Antietam in September) and Pennsylvania in 1863 (which culminated in the battle of Gettysburg in July), they might have interrupted the flow of coal from and through eastern Pennsylvania, the principal source of supply. As it was, the occurrence of an occasional coal miners' strike interfered with the coal supply and thus impaired the effectiveness of the blockade. Even with coaling stations near at hand, the commanders of the blockading squadrons sought to minimize the absence of blockading vessels from their stations—they instructed their ships' captains to obtain supplies when the Moon was fullest and blockade running was least likely.

The growth of the squadrons required growth of the facilities that supplied and supported them. Starting in 1862 the

USS *Vermont,* a ship-of-the-line sailing ship, served the South Atlantic Blockading Squadron in Port Royal as a stores ship for ordnance and other supplies, as a hospital ship and as a receiving ship for men awaiting their service assignments. By early 1863 the growth in the squadron's size caused the various functions aboard the *Vermont* to interfere with one another. The *Vermont*'s commander noted,

> The demands for ordnance materials are so numerous and so constant that to observe the usual precautions of the service in receiving and delivering such is to seriously interrupt and interfere with the daily wants of the ship herself. Not to observe them is to incur more or less of risk.

The commander proposed obtaining two additional hulks, one for powder and the other for loading shells. If two could not be obtained, he recommended at least that a separate hulk be obtained for the ordnance department.

The *Vermont* and her sister ship the USS *New Hampshire* (originally named the *Alabama*) were two of nine 74-gun ships-of-the-line authorized by Congress in 1816. The *Vermont* was laid down in the Boston Navy Yard in 1818, and the *Alabama* was laid down in the Portsmouth Navy Yard in 1819. Both vessels were completed in 1825. The *Vermont* remained in stocks until 1848 when she was launched. She was not commissioned but remained in ordinary until the need for her as a stores ship arose in 1862, when, in February, she was sent to Port Royal. The *Alabama* remained in stocks until the Civil War. She was renamed *New Hampshire* in 1863 and launched in April 1864. She was commissioned in May of the same year and replaced the *Vermont* in Port Royal in July. The *Vermont* sailed to New York where she served as a stores and receiving ship

5. **Facilitating Repairs.** Blockading was wearing duty, hard on the ships and hard on the men who served on them. From the late 18th century until the early 20th century—a period that embraces both the age of sail and the age of steam—the British experience was that 20 to 25 percent of any blockading force could be expected to be in port refitting. During the Civil War, the age of steam had begun, but the age of sail had not ended. Many shipmasters may have considered steam an auxiliary to sail, and ocean-going steamships carried masts, rigging and a full set of sails. Boilers and steam engines on some ships were expected to be run for limited periods of time. Even those vessels that crossed the ocean under full steam used the interval between crossings to clean, maintain and repair the boilers, engines and related equipment. The requirements of the blockade for constant vigilance and readiness meant that fires burned in boilers of the blockading vessels for months on end, with the result that the vessels were in frequent need of repair.

The vessel's engineers determined whether the boiler and engine remained in sufficiently good condition to be run at their rated steam pressure—a vessel operating at reduced pressure could not attain its full speed. The USS *Dawn*, for example, had a maximum allowed operating pressure of 45 pounds, and her usual steaming pressure was 35 pounds. Her flues were so badly worn—varying in thickness from one-eighth of an inch to one-thirty-second of an inch—that they could not be patched and might give out at any time, even with a boiler pressure reduced to 25 pounds.

The Federal Navy Department provided men and materials so that the blockading squadrons could make many repairs locally. These included a floating machine shop at Port Royal and land-based facilities at other places under Federal control. At the time of the Civil War, standardized manufacture had been applied to some products, but not to steam engines and boilers. Accordingly, there was no store of readymade replacement parts for some equipment; if an engine part was broken or badly worn, a replacement had to be made from scratch, which generally involved casting or forging a piece of iron and then machining and hand filing the replacement so that it was functionally identical to the original. Machine tools facilitated making repairs and replacements, but, even with that assistance, the work took time. As noted, the steam engines designed by Engineer-in-Chief Benjamin F. Isherwood anticipated the problems related to constant operation by emphasizing simplicity, reliability and robust construction. The Navy built relatively few of the hundreds of steamers that it brought into service during the war. Many more were built as civilian vessels and were converted to naval use, but these probably went into service with their original boilers and engines that were not designed with Chief Isherwood's priorities in mind.

The resources at the blockading squadrons limited the repairs that they could make, and if a squadron's machine shop could not do the work, the vessels needing the repair would have to go north. In the North, the aggregate demand for war materiel strained the available industrial capacity, so the vessels needing repairs had to wait. In November 1861 the Federal Navy Department sought a foundry to make a replacement shaft for the USS *Roanoke*, but all the establishments capable of doing the work were fully engaged. The department found a manufacturer to cast the shaft, but the manufacturer declined to do the machining needed to finish it. The Navy Department decided to finish the work in one of its own yards, but, since none of the Navy's machine tools could accommodate the shaft, a new machine tool had to be built. Even when facilities were capable of doing the work, vessels might have to wait their turn, or the work might take longer than expected; and even when the work got done, it might not have been done right. The USS *Flambeau* went north for repairs that were supposed to have taken between three and four weeks. When she returned to the squadron eight weeks later, she was still not ready to go into service.

As the fleet grew and the need for repairs increased, the Federal Navy Department expanded the repair facilities at the blockading squadrons. In April 1862 the Navy sent machines and 25 machinists to Key West. By August 1864 the shop had grown to 80 machinists. The repair facilities at Port Royal began as a single floating machine shop and grew to occupy seven of the hulks that housed a blacksmith shop, machine shops, a brass foundry, a boiler-making shop and a large furnace that was used to forge gratings, shafts and cranks up to 500 pounds. Admiral Farragut built a machine shop at Pensacola that included a large steam hammer. He also intended that the yard have a dry dock that would make it capable of making almost any repair that could be made in a northern port.

Admiral Dahlgren reported in January 1864 that he was at times short of the steam vessels he needed to perform the missions of his command due to their constant need for repair. He wrote, "The frequent disability of steamers even when new indicates something wrong, either in the engines or in the engineers.... The monitors are without first assistant engineers; in some the boilers begin to require careful watching." The likely fault lay in both the machines and the men, and the likely cause was the inadequate number of men with technical skills. The great demand for steam engines strained the capacity of northern industry, and the quality of the engines probably declined, which would increase the incidence of malfunction. At the same time, the Navy's need for engineers brought into service men who had less experience and skill, which could have resulted in less preventive care and more unintentional abuse of the machinery.

6. Providing Administrative Support. The British, mindful of the wearing effect of blockade duty on men, provided important blockades with two sets of flag officers. The Federal Navy Department did not follow this practice during the Civil War, although it divided the blockade into two and then four separate commands, and it instructed the squadron commanders to appoint an administrative staff including a fleet paymaster and a fleet engineer, who would "devote their whole time to systemizing their respective departments," and a fleet captain, who should be "your most systematic captain in the squadron." The fleet captain served as the chief of staff and sometimes was referred to as such, but administrative duties did not supplant combat duties. Over the course of three months in 1863, Admiral Dahlgren lost four fleet captains. Captain William Rogers Taylor fell ill and went north in July. Captain George W. Rodgers was killed in August while commanding the monitor *Catskill* during an attack on Fort Wagner. Captain Oscar C. Badger was wounded on the monitor *Weehawken* in an attack on Fort Moultrie in September. Lieutenant Samuel W.

Preston was captured during a boat attack on Fort Sumter also in September.

In March 1864 Admiral Lee wrote:

> Experience has proved to me the necessity of a chief of staff of the commander in chief of a fleet or squadron, whose duties shall be similar to those of the adjutant-general of our army in the field, and that he should have rank and pay commensurate with the character and extent of the duties he is called upon to perform.

The purpose of the fleet engineer was to keep the squadron at maximum strength by requiring that maintenance be performed to reduce the number of breakdowns and determining which repairs could be accomplished locally to avoid sending vessels away from station to northern ports. Squadron commanders also appointed the senior officer present to exercise local command at a place under blockade by several vessels.

7. **Deploying the Fleet.** Not all of the Navy's vessels were suitable for blockade duty—some vessels were too large and thus had too deep a draft for service in shallow coastal waters. Nor could all the suitable vessels be committed to the blockade inasmuch as the Navy had other duties to perform—on the Pacific Coast, in foreign waters and in cooperation with the Federal Armies. The Navy cooperated extensively with the Federal Army in operations along western rivers, including the capture of Fort Henry and Fort Donelson in February 1862, the capture of Island No. 10 in April 1862 and General Grant's campaign against Vicksburg, Mississippi, that resulted in the surrender of that city in July 1863. River operations continued

through the end of the war. In the east, the Navy supported General McClellan's peninsular campaign in 1862 that started with a landing at Yorktown, Virginia, and General Grant's 1864 overland campaign that moved south of Richmond and established a siege of Petersburg.

The Confederates launched cruisers to prey upon Federal commercial shipping, which pulled some Federal vessels away from the blockade. In May 1861 Secretary Welles created the West India Squadron to hunt cruisers and protect Federal commerce, but he instructed the squadron's commander that its mission was "secondary to the great object of the blockade." Secretary Welles discontinued the West India Squadron in August 1861 but reestablished it in September 1862 in response to the operations of the cruisers CSS *Florida* and CSS *Alabama*.

In addition to imposing the blockade, the blockading squadrons conducted raids on Confederate positions and facilities along the coast. These included cutting out operations, in which Federal forces raided a Confederate harbor for the purpose of seizing or destroying the vessels within, and the destruction of salt works located at or near the coast. In addition, by the middle of 1862, the South Atlantic Blockading Squadron, in some instances assisted by the Federal Army, had taken control of the coast of Georgia and a substantial portion of the coast of South Carolina and the Atlantic Coast of Florida.

The composition of the Confederate Navy changed over time, which modified the Federal Navy's responsibilities. At the beginning of the war, the Confederates possessed a Navy but lacked a fleet. In his report on the Confederate Navy Department on April 26, 1861, Secretary Stephen R. Mallory reported that the Navy possessed three vessels—the *Sumter* and the *McRae*, which were being fitted out as cruisers, and the *Star of the West*, which was useful only as a transport. Despite the difficulties that the Confederates encountered in acquiring vessels, the picture had changed by 1864 when Secretary Mallory reported that the Confederate Navy possessed the following vessels at the following stations:

- James River in Virginia: three ironclads and seven other vessels in commission; two ironclads and four torpedo boats under construction.
- Inland waters of North Carolina: two ironclads in commission.
- Cape Fear River in North Carolina: two ironclads, one floating battery and one gunboat in commission; two torpedo boats under construction.
- Charleston Harbor: three ironclads in commission; three ironclads, one gunboat and two torpedo boats under construction.
- Savannah River: one ironclad, one ironclad floating battery and two other vessels in commission; three ironclads and one gunboat under construction.
- Mobile Harbor: one ironclad, two ironclad floating batteries and four other vessels in commission; four ironclads and one side-wheeler under construction
- Red River: one ironclad
- St. Marks River, Florida: one gunboat

The report also listed four cruisers that were at sea or undergoing repairs. Although tiny in comparison to the 559 vessels, including 71 ironclads, that the Federals had at that time, the

Confederate Navy, especially its ironclads, represented a force that posed a threat to the Federals.

A review of the disposition of forces in each of the blockading squadrons in 1864—roughly the same time as Secretary Mallory's report—shows both the way in which the Federal Navy Department allocated resources to the squadrons in light of the circumstances presented in the areas to which they were assigned and the way that the squadron commanders allocated these resources to meet the challenges that they faced locally. The monthly reports submitted by the squadron commanders to the department included service and supply vessels and vessels undergoing repair locally as well as in northern ports.

The North Atlantic Blockading Squadron, responsible for the coast of Virginia and North Carolina, consisted of 114 vessels. By mid-June 1864 the Federal Army of the Potomac had advanced through Virginia to south of Petersburg, where the Federals and the Confederates opposed each other in trenches that were miles long. Supplies for the Federal Army came by water and were shipped to the front by rail. The squadron had 33 vessels on the rivers that ran through the lines. The advanced group of 12 vessels were just behind the Federal lines and included three wooden warships, the admiral's flagship, four ironclads and four tugboats that did double duty as torpedo boats and tenders to the ironclads. An additional seven vessels patrolled nearby waters. The squadron's other major concentration consisted of 25 vessels on blockade of Wilmington. The squadron stationed six vessels off the Roanoke River, where the ironclad CSS *Albemarle* operated.

The South Atlantic Blockading Squadron was responsible for the coast of South Carolina, Georgia and Florida as far south as Cape Canaveral. In June 1864 its force consisted of 80 vessels, and the largest group of these blockaded Charleston—23 vessels including six ironclads. The squadron had stationed another 35 vessels to patrol off 19 different points along the coast.

The East Gulf Blockading Squadron was the smallest squadron, and, in June 1864, it consisted of 35 vessels. The squadron was responsible for the Florida coast from Cape Canaveral to just east of Pensacola as well as Cuba and the Bahamas. The squadron stationed 16 vessels to patrol 11 points on the coast, and 4 more sailing vessels cruised along portions of the coast. An additional three vessels cruised the Gulf of Mexico and the coasts of Cuba and the Bahamas.

The West Gulf Blockading Squadron was responsible for the coast from Pensacola, Florida, to the Rio Grande. In June 1864 it consisted of 74 vessels, and it was moving vessels from the Mississippi River, where Admiral Farragut had been operating in support of Federal Armies inland, back into the Gulf of Mexico. The squadron had 13 vessels on blockade at Mobile and 7 at Galveston, and it had placed 19 vessels at nine other stations along the coast.

As these descriptions indicate, the squadrons allocated most of their forces to places where they were needed for cooperation with the Federal Armies or where the threat of blockade running was the highest. The places where the Federal Armies operated changed over time. The places where the threat of blockade running was high did not change except as a result of the blockade becoming more effective locally or the Federals capturing a Confederate harbor. The squadrons allocated small forces to deter blockade running in places where it was possible but less likely to occur.

6. Ramparts, Raiders and Rams

The state of Confederate supply reflected the relevance and the effectiveness of the Federal blockade. As noted, at the outset of the Civil War, the Confederates did not have on hand the resources necessary to fight even a limited war, and therefore the blockade could not be irrelevant. The steps that the Confederates took to develop the capacity to produce war supplies domestically and to allocate and ration their resources also were responses to the blockade, but they were tied up with various other efforts the Confederates made to marshal their war resources generally. These efforts at marshalling resources comprise a different history of which the blockade is only a part rather than the focus, and, accordingly, the Confederates' economic and industrial strategies are noted in this book with little further discussion.

The Confederates adopted various military strategies to counter the threat posed by the blockade. These included (1) defending Confederate ports to resist the Federal blockade and to encourage and assist blockade runners; (2) commissioning armed vessels—privateers and Confederate cruisers—to prey on Federal commerce; (3) launching Confederate ironclads to challenge Federal naval forces and, if possible, to bring the war to the Federals; and (4) developing other novel weapons to counter the Federal naval power. The Confederates' diplomatic strategies related to the blockade are addressed in Chapter Seven.

1. **Coastal Defense.** Initially, the Confederates responded to the war and the blockade by attempting to defend all points along their coasts and borders. Especially on the coasts, this approach spread available Confederate resources too thin to be effective at any given point, while the Federal forces afloat could move faster than Confederate forces on land and thus were able to concentrate their power at a point of their choosing before the Confederates could assemble an effective defense. Early in the war, the Federals captured Hatteras Inlet on the outer banks of North Carolina (August 28-29, 1861) and Port Royal Sound in South Carolina (November 7, 1861). In each case, the principal Confederate defenses were two small earthen forts. Federal warships brought overwhelming firepower to bear and put the defenders to flight.

After the fall of Fort Sumter, Federal forces retained possession of several coastal forts in Confederate territory. These included Fort Monroe on the Chesapeake Bay in Virginia; Fort Pickens on Pensacola Bay in Florida; Fort Taylor near the southern tip of Key West in Florida; and Fort Jefferson on Dry Tortugas in the lower Florida Keys. Fort Monroe helped the Federals assert control over the Chesapeake Bay, while the toeholds at Fort Taylor and Fort Pickens became coastal stations that supported the Federal blockade.

The Confederates revised their strategy—not as a national policy so much as the common pattern of actions taken by local commanders—to concentrate their forces and defense of a limited number of coastal sites. Elsewhere the Confederates stationed their forces inland, beyond the reach of the Federal Navy's large guns, and from where they could be sent to the places at which Federal forces landed. During the course of the war, this defense was weakened progressively as many of the troops assigned to the coast were reassigned to Confederate Armies engaged in combat inland.

Upon seceding from the Union, state forces seized most of the Federal facilities and supplies within their borders. Upon the formation of the Confederacy, the national government acquired control of many of these resources, the forts and larger armaments in particular. Throughout the war, however, state authorities acquired and possessed arms and other supplies that they used for state rather than national purposes, which arguably diminished or dispersed the Confederates' national military potential.

The principal assets that the Confederates acquired to defend against the Federal blockade were, of course, the stone and masonry coastal forts built before the war by the United States government. The Confederates supplemented these with earthen forts and batteries to the extent that manpower to construct and garrison them and guns to arm them were available. This concentration of forces increased the firepower that could be directed against the Federals and also could provide mutual covering fire to make individual fortifications less vulnerable to attack. Charleston provided the best example of this. Anticipating a naval assault through the harbor, the

Confederate commander arranged the guns of his land-based batteries to create three interlocking circles of fire through which an attacking naval force would have to pass to reach the city. Floating and underwater obstructions, mines and torpedo boats supplemented the harbor defenses.

At the major Confederate ports, earthen works were generally supplementary to existing stone and masonry forts. In one instance, however, an earthen fort became the more prominent defense facility. The eastern edge of the New Inlet, one of the two approaches to Wilmington, North Carolina, through the Cape Fear River, was originally called Federal Point, but, after secession, the Confederates called it Confederate Point. On this strip of beach, the Confederates originally built several individual batteries. But over time they consolidated these in an enormous earthen fort they called Fort Fisher. The sea face was about a mile long and rose about 12 feet above the water. The end nearest to the New Inlet terminated in the Mound Battery that was 30 feet high. At the opposite end, the land face ran perpendicular to the sea face and consisted of a line of 15 mounds each about 32 feet high and stretching a distance of 1,800 feet. The sea face mounted 22 guns, and the land face mounted 25.

2. Privateers and Cruisers. For a nation at war, the enemy's merchant vessels at sea were legitimate targets—a form of warfare called *guerre de course*—and, through the middle of the 19th century, such vessels were captured by the combatant's own commissioned naval vessels ("cruisers") or by private vessels ("privateers") that carried a government commission ("letters of marque"). Under international law, the capture of an enemy's merchant vessel by either a cruiser or a privateer needed

to be confirmed by a court of law, referred to as a prize court, duly constituted within the combatant's home country or the territory of an ally. Adjudication extinguished the property interest of the original owners; the purchaser of a prize sold without an adjudication assumed a substantial risk that the vessel would be reclaimed by its prior owners.

The possession of letters of marque and the adjudication of captures by a prize court constituted the veneer that separated privateers—lawful vessels of war—from pirates. The officers and crew of a cruiser or a privateer participated in a portion of the financial gain that resulted from the sale of the captured vessel and its cargo. Whereas cruisers were expected to attack the vessels of the enemy's navy, privateers seldom did so because of the likelihood that a naval vessel could possess superior arms and because the financial rewards of the potential capture did not justify the risk. The Confederates tried to create an economic incentive for their privateers to attack Federal Navy vessels by passing a statute that provided a bounty of $20 for each person aboard any armed Federal ship that was burned, sunk or destroyed by a privateer that was of equal or inferior strength. The statute assumed that a privateer of superior strength could and would effect a capture, and the statute provided an economic incentive for the humane treatment of enemy combatants by providing a bounty of $25 for each prisoner taken.

While many seafaring nations authorized privateering in previous times, by the mid-19th century many developed nations decided it was a thing of the past. Although in 1856 an international convention known as the Declaration of Paris abolished privateering, among other things, the United States declined to sign because it had relied upon privateers to supplement its traditionally small Navy. When the Civil War broke out, the Federals possessed a great merchant marine and a Navy that was tiny in comparison to the great naval powers, such as Britain and France. Nonetheless, the Federal Navy constituted a major naval force in comparison to the Confederates who had no Navy to speak of. When the Confederates elected to use privateers, President Lincoln's government offered to sign the Declaration of Paris. The British government agreed provided that acceptance by the United States did not apply to current hostilities with the Confederates, which negated the intended effect of the Federals' offer.

Early in the war, the Federal government captured some privateers and prosecuted the officers and crew as pirates, and, when they were sentenced to hang, the Confederates threatened to hang an equal number of captured Federal officers. The Federal government bowed to the threat and thereafter treated the captured privateers' men as prisoners of war. Although successful in their brinkmanship, the Confederates gained very little by it. In the earliest days of the Civil War, Confederate privateers had ready access to Confederate prize courts, but, as more Federal vessels joined the blockade, not only did privateering become riskier but also bringing the prize to a Confederate court became more difficult. When Britain and France declared their neutrality, they refused to permit Confederate privateers to use their courts to adjudicate the captures. Most potential buyers were unwilling to accept the risk of buying an unadjudicated prize, which made privateering unprofitable. With increased riskiness and reduced profit potential, Confederate privateering petered out. The Confederates issued at least 57 letters of marque (as compared to over 500 in the War of 1812) and launched 27 or more privateers that made a total of 27 or more captures during the course of the war. All but three captures occurred from May to October 1861. The others took place during January and February 1863 and were made by a single privateer, the *Retribution*, that operated out of Galveston, Texas, which was in Confederate hands.

During 1862 and 1863 when the Confederate cruisers were actively capturing and destroying merchant ships flying the United States flag, and the Federal Navy had not been able to stop them, the idea of granting letters of marque to Federal privateers to supplement the strength and reach of the Federal Navy gained popularity. Indeed, Congress passed a law authorizing the issuance of letters of marque. Although the hope was that the Federal privateers would hunt the cruisers, the reality was that the Federal privateers were likely to hunt blockade runners, which was less dangerous and offered a greater economic reward. The Federals did not grant any letters of marque—the one applicant did not express any interest in pursuing the Confederate cruisers.

Unlike a privateer, a cruiser was a government vessel that sought principally to inflict harm upon the enemy whether by attacking government vessels or capturing merchant vessels. Although the members of a cruiser's crew would share in prize money from captures, taking prizes weakened the cruiser because a portion of its crew had to sail the captured prize back to a friendly port. Accordingly, some cruisers destroyed captured vessels after making provision for the safety of the captured crew. Destroying the enemy's ships caused as much damage as a capture, and the doctrine of necessity permitted destruction when taking the captured vessel to a prize court was either impossible or highly inconvenient. Some cruiser captains bonded their prizes: they released the captured vessel upon the owner's written promise to pay the Confederacy the value of the ship and its cargo at the end of the war, a practice recognized under international law.

The Confederates sought cruisers both in their own waters and overseas. They purchased the steamship *Havana* (or *Habana*) in New Orleans and commissioned her as the CSS *Sumter*. She ran the blockade to sea on June 30, 1861, and began preying upon Federal merchantmen. Stephen R. Mallory, the Confederate secretary of the Navy, sent Commander James D. Bulloch, formerly of the United States Navy, to Britain to buy or build steam-powered ships to serve as cruisers. International law required neutral governments, and their citizens, to refrain from selling warships to the belligerents. Federal agents in Britain and France monitored the Confederate activities there and sought to interfere with Confederates' acquisition of cruisers.

Purchase was the preferred Confederate means of acquiring a potential cruiser because buying a merchant vessel appeared innocent, and, if conducted through a third party, the transaction could be completed before the Federals or the neutral government became aware of it. Once acquired, the vessel would be armed and commissioned as a cruiser. The difficulty with the purchase strategy was that few of the vessels available possessed the qualities needed in a cruiser. Building cruisers increased the risk of discovery since construction took a substantial amount of time, construction of a large vessel was impossible to conceal and the character of the vessel might be guessed from her appearance—a purpose-built cruiser, although not armed, had attributes that distinguished her from a merchant vessel, such as a magazine, gun ports, racks on the deck for holding shot and canister and the lack of a substantial hold for cargo. Moreover, Confederate agents visiting the shipyard to monitor construction was evidence of ownership.

Unable to find any suitable vessels for purchase, Commander Bulloch contracted to have British shipyards build two vessels that became the cruisers CSS *Florida* and CSS *Alabama*. The British government initially took a casual attitude toward the Confederate activities, but it became more aggressive after the Confederates spirited the *Florida* and the *Alabama* out of Britain and they began sinking Federal commercial ships.

One of the sharpest continuing points of contention between the British and the Federals involved the construction of Confederate cruisers in British shipyards. International law required each neutral nation to abstain from rendering assistance to either belligerent's war-making efforts and abilities, the possibility of giving equal war-making assistance to both being dismissed as absurd. The limits on official government action of a neutral nation in support of either belligerent were summed up in two principles:

> 1. To give no assistance where there is no previous stipulation to give it; nor voluntarily to furnish troops, arms, ammunition, or any thing of direct use in war....
>
> 2. In whatever does not relate to the war, the neutral must not refuse to one of the parties

> merely because he is at war with the other, what she grants to that other.

Similarly, the obligation of neutrality forbade a nation from permitting a belligerent to use neutral territory as a base from which to attack the enemy.

The actions of private citizens within the neutral nation could constitute a violation of neutrality as well. Arming and equipping vessels and enlisting men in a neutral port to enable one belligerent to attack another belligerent upon the high seas was contrary to the obligations of the neutral power. By contrast, the private sale of munitions to a belligerent for shipment to the belligerent's territory was not a violation of neutrality. Britain's Foreign Enlistment Act was a codification of its neutral obligations under international law. In addition to prohibiting British citizens from enlisting in the military service of a foreign government, the act also provided that no person may "equip, furnish, fit out, or arm" a ship or vessel to be used by a foreign government "to cruise or commit hostilities against" any other foreign government with which Britain was not at war. Commander Bulloch, the Confederate Navy Department's principal representative in Britain, obtained a solicitor's opinion that simply building a vessel did not violate the act so long as she was not armed when she left British waters. Accordingly, each vessel built in Britain for Confederate service sailed unarmed to a location outside of British jurisdiction where she was equipped as a cruiser—other than the *Florida* that was fitted out in the Bahamas. At that point, the Confederates acquired ownership and commissioned the vessel into Confederate service.

As the depredations of the British-built Confederate cruisers became known, the British government became more sensitive to its responsibilities and risks. On March 28, 1863, Lord Russell, Britain's foreign minister, wrote to Lord Lyons, Britain's minister to the United States:

> The outcry in America about the *Oreto* [as the *Florida* was named initially] and the *Alabama* is much exaggerated, but I must feel that her roaming the ocean with English guns and English sailors to burn, sink and destroy ships of a friendly nation, is a scandal and a reproach. I don't know very well what we can do.

On April 24, 1863, Lord Lyons responded, saying that the Americans were "drifting into a war" with Britain over the cruisers. The British government seized another prospective Confederate cruiser, the *Alexandra*, but a British court ruled that her construction did not violate the Foreign Enlistment Act. The notoriety surrounding the *Alexandra* caused her builder to sell her to a third party, and she became a blockade runner. Confederate shipbuilding in France started only after Britain stepped up its enforcement actions—France, in the person of the Emperor Louis Napoleon, gave unofficial encouragement to the Confederate construction program. When the Federals got wind that vessels were being built for the Confederates, the French government prevented their delivery.

During the course of the war, the Confederates commissioned nine vessels as cruisers that captured over 200 merchant ships, most of which were destroyed, and the cruisers diverted some Federal warships from the blockade in an attempt to hunt them down. (See Table 6-1.) The Confederates contracted for an additional 11 cruisers in British and French shipyards that were not commissioned into Confederate service; one sailed to Charleston, but a Federal blockader damaged her, and she was abandoned before being fitted out as a cruiser; and of the rest, either the government prevented transfer to the Confederates or the vessels were not finished before the war ended. (See Table 6-2.) Only one Confederate cruiser fought with Federal Navy vessels. On January 11, 1863, the CSS *Alabama* engaged and sank the USS *Hatteras* off the Texas coast near Galveston, and, on June 19, 1864, the *Alabama* engaged and was sunk by the USS *Kearsarge* off the coast of France near Cherbourg.

TABLE 6-1: CONFEDERATE CRUISERS COMMISSIONED					
Cruiser	Commissioned	Length of Cruise in Months	Captures	Origin	Disposition
Sumter	June 1861	19	18	Purchased in New Orleans and converted to cruiser.	Blockaded in Gibraltar and sold.
Nashville	October 1861	6	2	Seized in Charleston and converted to cruiser.	Sold as blockade runner.
Florida and tenders	August 1862	27	37	Built as a cruiser in Britain.	Captured by USS *Wachusett* while anchored in Bahia, Brazil, in violation of Brazilian neutrality.

Cruiser	Commissioned	Length of Cruise in Months	Captures	Origin	Disposition
Alabama and tender	August 1862	11	60	Built as a cruiser in Britain.	Sunk by USS Kearsarge off Cherbourg, France.
Georgia	April 1863	17	9	Built as a cruiser in Britain [?].	Captured by USS Niagara off Portugal.
Rappahannock	November 1863	0	0	British corvette purchased as cruiser.	Escaped to Calais where detained by French government.
Tallahassee	July 1864	2	32	Built as a cruiser in Britain.	Renamed Olustee and continued service as cruiser.
Chickamauga	September 1864	6	7	Former blockade runner. Purchased in Wilmington and converted to cruiser.	Burned to prevent capture in Fayetteville, North Carolina, February 1865.
Shenandoah	October 1864	11	38	Built in Britain, purchased in Wilmington and converted to cruiser.	Surrendered to British authorities in November 1865.
Olustee	August 1864	2	6	CSS Tallahassee renamed.	Converted to blockade runner.

Source: P.H. Silverstone, *Civil War Navies*.

TABLE 6-2: CONFEDERATE CRUISERS THAT DID NOT SEE SERVICE

Cruiser	Launched	Origin	Disposition
Georgiana	December 1862	Built as a cruiser in Britain.	Damaged by USS *Wissahickon* and abandoned off Charleston where she was to be fitted out as a cruiser.
Alexandra	March 1863	Built as a cruiser in Britain.	Seized by British government April 1863.
Texas	October 1863	Built as a cruiser in Britain.	Seized by British government December 1863.
Louisiana	May 1864	Built as a cruiser in France.	Embargoed by French government February 1864.
Mississippi	1864	Built as a cruiser in France.	Embargoed by French government February 1864.
Texas	1864	Built as a cruiser in France.	Embargoed by French government February 1864.
Georgia	1864	Built as a cruiser in France.	Embargoed by French government February 1864.
Ajax	December 1864	Built as a cruiser in Britain.	Completed too late for Confederate service.
Hercules	December 1864	Built as a cruiser in Britain.	Completed too late for Confederate service.
Adventure	1865	Built as a cruiser in Britain.	Completed too late for Confederate service.
Enterprise	1865	Built as a cruiser in Britain.	Completed too late for Confederate service.

Source: P.H. Silverstone, *Civil War Navies*

The ability to strike the enemy directly—even if by proxy in the form of its merchant and whaling vessels—was almost irresistible to the Confederate Navy Department, but the cruisers did not substantially aid the Confederate war effort. They did not materially impair the flow of goods to and from the northern states. Unlike Britain during the world wars of the 20th century, the United States in the mid-19th century did not depend upon its overseas trade either to sustain its war-making power or for its survival as a nation. While the Confederate cruisers drew some Federal vessels away from the blockade, maintaining a strong blockade remained the Federals' first priority. The main effect of the cruisers' campaign was to drive American merchant ships to foreign ownership.

Indeed, commissioning cruisers proved somewhat counterproductive. Confederate supply was never plentiful, but, in 1864, with supply problems worsening, the Confederates nonetheless converted two blockade runners into cruisers that sailed from Wilmington—this not only reduced the number of blockade runners but also drew Federal attention to Wilmington, which was becoming increasingly important as a source of supply for General Lee's Army, a fact that drew General Lee's disapproval.

3. Ironclads. People knowledgeable about naval matters were aware in early 1861 of the potential power of armored ships. On May 17, 1861, Secretary Mallory ordered Lieutenant James H. North to proceed to Europe to attempt to buy the *Gloire* from France or, failing that, to have one or more armored ships built. France refused to sell the *Gloire*, and Lieutenant North found the British and French shipyards fully engaged building ironclads for their governments. Accordingly, the Confederates started their ironclad program with vessels built at home.

With a monopoly—or a local monopoly—of even a small number of ironclads, the smaller Confederate Navy could be the stronger power (even if only locally), and it might break the blockade and carry the war into northern ports. When the CSS *Virginia* made her initial entrance into Hampton Roads, the panicked Federal government shared this belief—it telegraphed urgent warnings to northern coastal cities and prepared to obstruct the channel of the Potomac River below Washington. Although the Confederates gained a head start in converting the steam frigate *Merrimack* into the ironclad *Virginia*, the Federals were able to build and launch the ironclad *Monitor*, which was at least a match for the *Virginia*, in sufficient time to limit the Confederate monopoly on ironclads in the Chesapeake Bay to one day.

All but one of the Confederate-built ironclads used the casement design on top of a wooden hull—a roof-like structure with slanted sides and a flat top covered most of the hull. The construction was principally wood, covered with layers of iron plate. A curtain of iron plates was mounted where the slanted sides of the casement met the hull and extended below the water line to protect the hull. Embrasures—openings through which cannon were fired—pierced the sides, front and back of the casement. The judgment of the men who commanded the Confederate ironclads was typically that they were not seaworthy and thus unsuited for venturing beyond protected coastal waters. The vessels also lacked sufficient motive power. The *Virginia* used the steam engine from the *Merrimack*, which, although only five years old, had been condemned by the United States Navy. Most of the other Confederate ironclads had steam engines that were scavenged from lighter vessels, so the ironclads, laden with tons of iron plate and heavy guns, moved sluggishly. Once the element of surprise had been lost, such cumbersome vessels would be unable to pursue and destroy faster warships.

The Confederates' efforts to build their ironclads—in the context of waging a major war—overtaxed their limited industrial capacity. As noted above, the Confederates possessed limited facilities for building ships, and their principal facilities that predated the war—located in Norfolk, Pensacola and New Orleans—were lost by May 1862. The Confederates undertook the construction of 32 ironclad vessels in the coastal states and commissioned 19 of them into service. Without setting any priority among the projects, they all suffered delays. Initially the South lacked the capacity to roll iron plates, and, although the Confederates built rolling mills, the supply of iron was not sufficient to meet all the demands for iron products, including cannon, cannonballs, shells and railroad rails in addition to armor plate. Limited transportation capacity slowed the movement of the iron plate to the construction sites. The Confederates also lacked manpower generally as well as skilled manpower, such as mechanics, machinists and engineers. Secretary Mallory tried to hasten construction and remove impediments, but the needs of the Confederate Army generally took priority over the Confederate Navy in access to resources.

While the Confederate ironclads enjoyed a few limited victories in coastal waters, they generally suffered crushing defeats:

- When the CSS *Virginia* entered Hampton Roads on March 8, 1862, she found that three Federal warships had run aground and were unable to flee. The *Virginia* rammed one and set another on fire with hot shot. She would have finished off the third the next day had not the *Monitor* arrived and fought the *Virginia* until the battle broke off.

- When Flag Officer Farragut sailed up the Mississippi River on his way to New Orleans on April 24, 1862, the Confederate forces opposing him possessed three ironclads, and the Federals had none. The first, CSS *Manassas,* mounted a single cannon and possessed inadequate iron cladding and was among the Confederate vessels defending the river. The *Manassas* rammed several Federal vessels, then ran aground and was burned. The second, *Louisiana*, remained incomplete (and was not commissioned) and could barely move under her own power. She was lashed to the riverbank and used as floating battery. The third, the *Mississippi,* remained in the shipyard when the attack occurred.

- On January 31, 1863, the Confederate ironclads CSS *Chicora* and CSS *Palmetto State* came out of Charleston Harbor and attacked the blockading vessels. The *Palmetto State* rammed the USS *Mercedita,* which surrendered, and the other blockading vessels nearby withdrew. The USS *Housatonic* advanced on the ironclads, and they returned to Charleston and did not inflict any further damage during the war. The Confederates sank both vessels when they evacuated the city in 1865.

- The steamer *Fingal* had brought war supplies into Savannah in November 1861. The Confederates converted her into the ironclad CSS *Atlanta,* and, on the morning of June 17, 1863, the *Atlanta* advanced toward Wassaw Sound where two monitors were waiting. The

Atlanta opened fire. The monitors advanced until they were within 300 yards of the *Atlanta*, then the lead monitor opened fire, striking the *Atlanta* four times. The *Atlanta* surrendered.

- The Confederates possessed a local monopoly in ironclads when the CSS *Albemarle* appeared near Albemarle Sound in North Carolina in April 1864. The Federal vessels present attempted to ram and surround her, but the *Albemarle* was able to withdraw. The Federals sank the *Albemarle* at her dock with a nighttime torpedo boat attack.

- When Admiral Farragut's warships entered Mobile Bay on August 5, 1864, they knew that the ironclad CSS *Tennessee* was waiting for them. The Federal force included two monitors and two double-turreted river monitors, but one of the monitors hit a torpedo and sank before entering the bay. Most of the remaining Federal vessels surrounded the *Tennessee* and fired their guns into her at close range until she surrendered.

As the war entered its second year, Commander Bulloch found British shipyards willing to undertake the construction of an armored ship, although he was told that completion of a seagoing ironclad would require at least a year. Commander Bulloch proposed that the iron plates and parts, difficult to come by in the Confederacy, be manufactured in Britain and shipped to the Confederacy for assembly. This would have been more clandestine than constructing complete vessels in Britain and might have circumvented difficulty under the British neutrality law, but the Confederate Navy Department did not take up Commander Bulloch's suggestion. Rather, the Confederates contracted to build a pair of ironclads in Britain, but the British government became aware of the project and purchased the vessels for the Royal Navy. The Confederates also arranged for the construction of a large ironclad in Scotland and two smaller ironclads in France. The contract for the Scottish boat was canceled in 1863. Information about the construction in France was leaked to Federal agents, and, when confronted, the French embargoed both vessels. One of them was sold to Denmark, and, while the vessel was in Denmark for inspection, the builder arranged for the Confederates to purchase her. Commissioned as the CSS *Stonewall*, she entered Confederate service too late to see any action and was surrendered to Spanish authorities in Havana.

4. Torpedoes, Torpedo Boats and Submarines. The Confederates sought novel weapons, in addition to ironclads, in an attempt to offset the expected superiority of Federal naval forces. One such weapon was the torpedo, an explosive device that later generations referred to as an undersea mine. Such devices had been used by the Americans during the Revolutionary War and by the Russians during the Crimean War, although without much success. Designs varied, but, in essence, a torpedo was a canister of gunpowder that detonated near the enemy's vessels. Some torpedoes floated and either were set adrift or were anchored in a fixed location on or under the surface. One of the earliest torpedoes was found floating in the Potomac River—a length of fuse had been lit as it was set adrift. Later floating torpedoes had percussion fuses, and other torpedoes, mounted on wooden frames beneath the water, were detonated by percussion fuses. Still

other torpedoes were placed on the bottom in relatively shallow water and detonated with an electric charge by an operator on shore when a ship passed over the torpedo.

Local defense forces designed and built the earliest torpedoes used during the Civil War, and they achieved only limited success. When the Federals attacked Fort Sumter with their ironclads in April 1863, they feared that the Confederate defenses included torpedoes, and they reported that a torpedo detonated near the USS *Weehawken,* although it did no damage—Confederate accounts differ as to whether any torpedoes had been deployed in Charleston Harbor at that time. According to the statement of a Confederate deserter, the *New Ironsides* took a position in Charleston Harbor that was directly over a large torpedo—constructed from a boiler containing 12,000 pounds of gunpowder—but the torpedo's electric firing mechanism failed to function or had been sabotaged.

Secretary Welles discounted the threat posed by torpedoes when writing to one of his squadron commanders in November 1863: "The torpedoes hitherto encountered during the war have not proved dangerous or serious preventatives to naval operations." To that point in time, through two-and-a-half years of war, torpedoes had damaged nine Federal vessels and sunk three. In the interim, the Confederates organized the Confederate States Submarine Battery Service and a Torpedo Bureau based in Richmond that promoted inventions, conducted experiments and sent emissaries to Europe to collect information and materials. ("Submarine battery" referred to torpedoes, detonated by an electric charge from the land, that protected the navigable channels. The Torpedo Bureau dealt with all kinds of submarine torpedoes and landmines.)

Confederate torpedoes caused more extensive injury during 1864, sinking 19 Federal vessels and damaging 3 others. Torpedoes were relatively inexpensive and eventually proved to be effective weapons. Not only could they sink ships, but even the suspicion that torpedoes were present generally would curb aggressive Federal action. Yet torpedoes were unreliable. The lead ship in Admiral Farragut's flotilla turned wide as it pushed into Mobile Bay in August 1864. It did not pass through the ship channel under the guns of Fort Morgan as intended (and over a submarine battery) but through a field of contact torpedoes. (A Confederate report noted that the Federal vessels were so far from shore that the projectiles from boat howitzers mounted in their masts did not strike.) The monitor *Tecumseh* struck a mine and sank in under a minute. The rest of the flotilla proceeded through the torpedoes without further explosions, although the reports mention sounds that might have been detonators snapping. The slightest leak that admitted moisture into the torpedo neutralized the gunpowder. Seawater and marine life could corrode or impair the firing mechanisms.

The Federals tried different strategies to counter the torpedo threat. John Ericsson, who designed the *Monitor*, built a raft that fit onto the prow of a monitor and that was equipped with grappling hooks intended to drag for torpedoes. The raft also had an explosive charge suspended beneath it that was supposed to blast away underwater obstructions. (The commander of one of the monitors that participated in Admiral Du Pont's attack on Fort Sumter agreed to use the raft but refused to carry the explosive charge.) Some vessels had netting rigged from spars extended ahead of the bow like a cowcatcher on a locomotive. Although the USS *Ostego* on the Roanoke River in

North Carolina was equipped in this manner, two torpedoes exploded under her just as she was coming to anchor, and she sank in a few minutes, but without serious injury to her crew. Two more torpedoes were found in her nets and six more torpedoes were found nearby. A Federal naval officer wrote, "The enemy sent torpedoes of various designs at the monitors so frequently that it became necessary to surround them with a heavy torpedo-netting of ropes supported by spars projecting from their sides. The sailors called them hoop-skirts." In April 1865 the Federals undertook a major torpedo clearing operation in the James River. Twenty boats proceeded up the river in an angled line, dragging for torpedoes. Squads of armed men, protected by skirmishers, advanced along both banks of the river ahead of the boats and were instructed to cut any torpedo wires in two places.

The Confederates defended their harbors with physical obstructions as well as torpedoes, and they often used the two in combination. The obstructions served as the means to mount torpedoes, and they served as barriers to prevent the passage of Federal vessels. These barriers consisted of wooden pilings driven into the harbor bottom, floating barriers composed of ropes, barrels and wooden beams and hulks or wooden cribs laden with stone. The Confederates generally prevented the Federals from removing or dismantling the obstructions by covering them with their guns. Lengths of rope hanging from the obstructions were intended to entangle propellers and hold the vessels under the fire of Confederate cannon. Obstructions that incorporated barrels were intended to suggest the presence of torpedoes. While some obstructions were permanent, others were temporary or movable since Confederates encouraged the passage of blockade runners and their own vessels into these same waters, but inevitably the obstructions and torpedoes occasionally disabled or damaged some of these vessels.

The United States had employed a gunboat strategy during the Revolutionary War and at various times through the early days of the republic. Updated with steam, rifled guns and exploding shells, the Confederate gunboat strategy looked promising. The gunboats would be just large enough to provide a firing platform for two large rifled cannon, mounted on pivots. The 20 rifled cannon of 10 such gunboats would fire as much iron as a steam frigate like the USS *Niagara*, which was armed with 10 large smoothbore guns. In that rough algebra, the combatants would be evenly matched, but the Confederates expected their gunboats to have a number of advantages. First, their rifled cannon would have range and striking power that was superior to the smoothbore cannon on Federal warships—the Confederate gunboats could attack a Federal warship from beyond the effective range of its smoothbore guns. Second, the gunboats would be less vulnerable to enemy fire because they presented a much smaller target and would, therefore, be much harder to hit—the Niagara presented a target of 10,000 square feet while each gunboat was estimated to present a target of 40 square feet. Even with 10 gunboats participating in the attack, the disparity in aggregate target size was enormous. Third, if enemy fire struck a gunboat, the damage suffered would be confined to that one gunboat—the larger Federal warship concentrated all of its resources and fighting ability in a single unit, so any damage threatened the whole. An additional benefit of the gunboat strategy was economy—each gunboat was estimated to cost $10,000, while the cost of the *Niagara* was in excess of $1 million, so for the cost of one steam frigate, the Confederates could build 100 gunboats. In addition, because of their small size, gunboats did not need to be built in a shipyard—they could be built at any point along the water's edge. Wood for construction was plentiful. The Confederate Congress in December 1861 authorized the construction of 100 gunboats, but, in about a month's time, the gunboat project had run into difficulty from not being able to engage enough carpenters and artisans. Securing 100 steam engines and 200 rifled cannon also would have posed problems. Meanwhile, the battle between the USS *Monitor* and the CSS *Virginia* heightened the enthusiasm for ironclads. Construction started on 15 gunboats in Virginia and several more in North Carolina and Florida, but none was completed.

The torpedo boat was a more successful innovation that married the destructive capacity of the torpedo with the aggressive ability and budget-mindedness of the gunboat. The torpedo boats, sometimes referred to as "torpedo rams," were low, swift steamers that were armed with a torpedo mounted at the end of a spar or boom at the bow. The torpedo generally had a percussion fuse and was intended to detonate against the hull of a Federal warship at or below the waterline. The torpedo boats relied upon speed and stealth for attack and defense. To attack a Federal warship, the Confederates needed either one torpedo boat, one ironclad or a squadron of gunboats, and to build one torpedo boat took less material, labor, armor, armament and machinery than either an ironclad or a squadron of gunboats, and they needed fewer men to operate them. Further, a single torpedo, delivered below the waterline, was potentially more lethal than a single gunboat. Since the torpedo boats could operate alone, and, because they were small, they were suited to night operations when darkness made them more difficult to defend against.

On the other hand, the torpedo boat's small size and low freeboard put it at risk of swamping if waters were not smooth. The need to be close to the target in order to deal a lethal blow also put the torpedo boat at risk from the success of its own attack—a 60-pound charge of gunpowder suspended at the end of a 10- or 20-foot spar did not leave much margin for safety. Moreover, the torpedo boat needed to maintain some momentum toward its target in order to detonate the percussion fuse, which carried the attacker toward the detonating explosive charge.

In anticipation of a Federal ironclad attack upon Charleston in the spring of 1863, the Confederates assembled an assortment of oared boats—skiffs, canoes and cutters—each armed with a 60-pound torpedo mounted at the end of a 20-foot spar. The Confederates planned to attack the Federal ironclads if they passed Fort Sumter and entered the inner harbor. The conditions were not favorable to the torpedo boats—many of the boats were in poor condition; their crews were inexperienced; and if the Federals attacked during daylight, which was likely, the element of surprise would be reduced—but the Confederates hoped for success because they believed that the Federals would not expect such an attack. When the Federal ironclads attacked Fort Sumter, they did not advance into the inner harbor, so the torpedo boats did not go into action. The Confederates also planned two nighttime attacks on the ironclads with their oared torpedo boats, but these did not take place.

Nighttime attacks with steam-powered torpedo boats showed more promise. At one in the morning of August 21, 1863, a torpedo boat attacked the *New Ironsides* that was at anchor off Charleston. Lookouts spotted the torpedo boat just as it approached, but by then it was too close to be fired upon except with small arms. The torpedo boat had trouble steering and was unable on two tries to strike its torpedo against the hull with sufficient force to detonate it. A second attack on the *New Ironsides* on the evening of October 5, 1863, was more successful. Again, the torpedo boat, named the *David*, was not seen until it neared the *New Ironsides*. The hail from the *New Ironsides* was answered by a shotgun blast. Moments later an explosion erupted beneath the *New Ironsides* hull. The torpedo boat retreated and disappeared. The explosion threw up a column of water that swamped the *David* and extinguished the fire in her boiler, and the impact threw part of the iron ballast into the machinery. Two men, believing the boat was sinking, swam off and were captured. The pilot stayed with the boat, and the engineer, having been thrown overboard, returned, rekindled the fire and got the machinery running. The two escaped with the *David*. The *New Ironsides* did not sink, but the damage was severe enough to require her to be taken out of service for repairs.

The *New Ironsides*, the largest ironclad in the Federal fleet, was 232 feet long and over 57 feet wide, had a crew of 460 men and carried 20 large guns. The aptly named *David* was about 50 feet long, 6 feet across and described as cigar-shaped—a long narrow iron cylinder with a cone attached at each end—with a crew of four. A 10-foot spar at the bow held a torpedo—with 60 pounds of gunpowder—about 6 feet beneath the water. Ballast submerged the boat so that only about 12 feet of its upper surface was visible. After the attack, a cap was placed over the smokestack to prevent extinguishing the fire in the boiler.

The *David* attempted other attacks in the spring of 1864; the torpedo used to attack the USS *Memphis* failed to detonate; heavy swells that threatened to swamp the *David* frustrated the attack against the USS *Wabash*. Another torpedo boat attack, on April 8, 1864, severely damaged the USS *Minnesota* off Newport News, Virginia. As with their ironclads, the Confederates began construction of a number of torpedo boats in Richmond, Charleston and other points during 1863 and 1864, but only a few of these saw service. The paucity of Confederate domestic industrial capacity caused the Navy Department to seek vessels overseas. In April 1864 Secretary Mallory instructed Commander Bulloch to acquire and send 12 small marine engines, and two months later he ordered six complete torpedo boats. The request for engines did not reach London until August, and Commander Bulloch sent these starting in mid-October (at least one group of six was stopped by the blockade), and the first of the torpedo boats was ready for shipment in late January 1865, but, by that time blockade running into the Confederacy had all but ceased, and the war would last only several months more.

Most of the torpedo boat activity occurred around Charleston, but Federal commanders elsewhere became aware of the threat. The Confederate torpedo boat attacks did not succeed in sinking any Federal warships, although the Federals used a torpedo boat to destroy the CSS *Albemarle*, a successful Confederate ironclad that challenged Federal control of the inland waters of North Carolina.

In addition to surface craft and semisubmersibles like the *David*, the Confederates experimented with submersible torpedo boats that would be known as submarines. Moving under the water improved the stealth with which the boats could approach their targets, and a layer of water also made them more difficult to damage with conventional guns. The Confederates built several submarines, and several crews drowned in the process of testing them. The best known was the *H.L. Hunley,* which was built in Mobile and then shipped to Charleston. Built of iron, the *Hunley* was cigar shaped. Flooding tanks attached to the *Hunley* caused her to submerge, and forcing the water out of these tanks with a hand pump brought the boat to the surface. Eight men turning a hand crank spun the propeller to drive the boat forward, and a ninth man steered. A small cupola on the deck with glass inserts permitted enough vision for navigation. The air supply was limited to what was held within the vessel when the hatch was closed. On the evening of February 17, 1864, the *Hunley* attacked the USS *Housatonic,* a steam gunboat with 13 guns anchored off Charleston Harbor. Sailors aboard the *Housatonic* did not see the *Hunley* until she was about to strike. The *Hunley*'s torpedo—mounted on the end of a spar attached to the bow—breached the hull of the *Housatonic,* which sank rapidly. The *Hunley* backed away from the *Housatonic* but did not return to Charleston. The *Hunley* and the remains of her crew were recovered from Charleston Harbor in the year 2000.

The Federals countered the threat of Confederate floating torpedoes, torpedo boats and submarines by surrounding their vessels with log booms and submerged netting and keeping smaller vessels—steam tugs, cutters and scout boats—on patrol at night. They also employed searchlights and added torpedoes and torpedo boats to their arsenals. These measures were largely successful in deterring Confederate torpedo boat attacks, although they consumed resources and tempered Federal aggression

The Confederates' ambitions for building innovative naval weapons was constrained by both the limits of the technology then existing and their own limited industrial capacity, and, as a result, they started many more projects than they completed. In hindsight, this failure rate suggests that their more prudent course would have been to review the projects more thoroughly, to proceed with only those that had a high probability of success and to channel their limited resources into fewer projects. Confederate ironclads, torpedoes and torpedo boats were sufficiently effective that the threat posed by information that a construction project had started nearby put the Federals on their guard, but neither the threat nor the presence of these weapons drove the Federals off. Paradoxically, if the Confederates had limited the number of projects they undertook and brought those few to fruition, the aggregate threat that the Federals perceived might have been diminished.

In their efforts at obtaining naval weapons overseas, the Confederates succeeded in building and launching cruisers. They failed at building and launching ironclads, and, although the cruisers destroyed a number of Federal merchant vessels, they did not materially affect the course of the war. Hindsight suggests that the better course would have been to supplement the machinery and resources that were scarce in the Confederacy with those obtained from Britain and France for the purpose of reinforcing its coastal defenses. More and better ironclads, torpedoes and torpedo boats might have been more effective in countering the force that the Federals were capable of bringing to bear.

7. Jealously Guarded Prerogatives

Although the Federals would have preferred to view the Civil War entirely as a matter of domestic concern, various international issues arose throughout the war and threatened to alter its course. A number of these were integral to the war for control of the Confederate coast inasmuch as the oceans were highways of international commerce and the Federal blockade was attempting to interdict the commerce of the Confederates with foreign nations. The issues discussed in this chapter include the legal basis of the blockade, Britain's attempts to moderate the Federals' blockading tactics for the benefit of blockade runners, the Federals' efforts to prevent the products of Federal industry and agriculture from supplying the Confederates or assisting the blockade runners, the lack of neutrality-in-fact in the conduct of certain representatives of foreign nations, and the relations of the Federals and the Confederates with their neighbors Canada and Mexico.

Countries at peace deal with one another based upon self-interest that is leavened on occasion by small measures of altruism. Especially when countries' interests conflict, each country sees itself as being motivated by benign and enlightened self-interest, and they see all others as being motivated by malign and cynical self-interest. The tendency to feel righteous about one's own actions and beliefs, and to disparage the actions and beliefs of another, is as natural for countries as it is for individuals. Historians are no less subject to the same nationalistic myopia when they ascribe motivations to foreign governments. Point-of-view can skew the accuracy of one's perceptions as much as can self-interest. The surer indicators of historical truth reside in action and in refraining from taking action. The less certain indicators are the explanations given for one's own actions or failure to act or for someone else's actions or failure to act. Explanations for one's own behavior may be discounted (although not dismissed) as so much self-justification, just as the explanations for another's behavior may be discounted (although not dismissed) as evidencing paranoia or contempt.

The conditions for achieving victory were different for the Federals and the Confederates, and so too were their intentions when dealing with the other nations of the world. The Confederates possessed a near monopoly on the world's supply of cotton, which motivated Britain and France, the principal consumers of raw cotton, to involve themselves in the American Civil War. Having declared their political independence, the Confederates sought diplomatic recognition, and their representatives sought to impress upon the European powers that the Confederates asked only for recognition as opposed to military intervention. Several Confederate secretaries of state wrote that they believed that the international recognition of the Confederate states would be sufficient to end the fighting, although the basis of this belief is not clear in light of the Federals' determination to prosecute the war. Although recognition might well have been the first step along a path that led a foreign nation into a more direct involvement in the Civil War, at first blush it appeared to require less cost or commitment from a European power than any form of direct involvement and therefore seemed easier to ask for and easier to give.

International law stated that premature recognition of independence was a just cause for war. A cogent essay on the subject, published in Britain during the Civil War, compared the examples of Britain's recognition of the independence, on the one hand, of Spain's former colonies in South America and, on the other hand, of Greece, formerly part of Turkey's empire. In the case of the nations of South America, Spain had ceased to take active steps to assert dominion over her former colonies notwithstanding her assertions to the contrary. In the case of Greece, Britain and a group of other nations expressed their willingness to use force to defend Greek independence. Thus, while recognition of Greek independence, while possibly premature, may have given Turkey cause to declare war under international law, Britain's readiness to go to war did not thus negate Turkey's position under international law, but rather was an acceptance of the consequences that might ensue. In the Civil War, the Federals were clearly attempting to extinguish the Confederate claim of independence, although, given the difference in the belligerents' differing conditions of victory, the Federals had a far more difficult task. Although the Federals might threaten war against a foreign nation that recognized Confederate independence, any actions by the Federals to conduct a foreign war while also waging war against the Confederates would make an eventual Federal victory over the Confederates all the more unlikely.

In evaluating the various international incidents that are related to the contest for control of the Confederate coast, we must keep in mind the larger context of facts and events—both as between nations and within individual nations—that shaped the attitudes of the leaders of the nations involved. The attitudes of Britain, and to a lesser extent France, were not static but changed over time in response to changing facts and subsequent events. From an initially somewhat arrogant hostility to the Federals coupled with a desire to secure a supply of cotton, perhaps with an underlying desire to see the American nation permanently sundered, those attitudes changed to a more cautious wariness that was mindful of the obligations of neutral powers. Consider the following:

- Immediately before the Civil War, the United States Army consisted of about 16,000 men, and its Navy had 42 vessels in commission, scant numbers compared with European armies and navies. During the war, the Federal Army grew to over one million men, and its Navy had hundreds of ships in service, including dozens of ironclads, which clearly changed other nations' perceptions of Federal military power.
- On November 8, 1861, Captain Charles Wilkes, commanding the USS *San Jacinto*, stopped the British mail packet *Trent* at sea and removed from her two Confederate emissaries on their way to Europe and a packet of Confederate dispatches. Captain Wilkes arguably had the right to bring the *Trent* into port

for adjudication as a prize, but removing the emissaries was contrary to international law. Although the incident made Captain Wilkes a national hero, Britain threatened war if the emissaries were not released. The Lincoln government acquiesced. President Lincoln is supposed to have said, "One war at a time."

- Although the Federal government backed down when Britain threatened war over the *Trent* incident, the experience of Britain's planning for a possible war showed the difficulty of projecting military power across the Atlantic, the unsuitability of the British fleet for blockading the Federal coast and the vulnerability of Britain's Canadian possessions. In her preparation for a possible war with the Federals, Britain sent troops to reinforce Canada. They arrived in North America in early January 1862, and Secretary Seward granted permission for the British troops to land in Portland, Maine, for transit to Canada. Although Secretary Seward was criticized at home for his act of comity during the crisis, his gesture doubtless had a sobering effect upon the British government.

- France invaded Mexico in late 1861 with the intent of replacing her republican government with a monarchy. But for the fact that the Civil War was occupying the Federals' attention, the United States probably would have asserted the Monroe Doctrine to assist Mexico. Throughout the Civil War, France maintained an Army in Mexico to prop up the new monarchy she had installed there, and this continuing commitment made France generally unwilling to provide partisan support to the Confederates without full British participation.

- Britain's failure to prevent the sailing of two cruisers built for the Confederacy in private British shipyards—the CSS *Florida* and the CSS *Alabama*—appeared to be a flagrant disregard of her neutral obligations as those cruisers captured and sank dozens of Federal merchant vessels. A somewhat more circumspect Britain prevented any further warships from being launched for the Confederates. The French government condoned the secret construction of warships in France for the Confederates, and the French government intervened to prevent the ships from sailing only after Federal agents made the construction program public. One French-built ironclad escaped and entered Confederate service under the name CSS *Stonewall* but too late to see any action in the war.
- Britain actively considered, but ultimately rejected, a French proposal to recognize the Confederacy and intervene in the Civil War if the Federals refused to agree to a ceasefire and arbitration. Although the proposal might have restored the cotton supply and reduced unemployment in the British and French cotton mills, it also could have brought war with the Federals, the cost of which would have exceeded the amount being paid to those who were put out of work in cotton mills and related industries. One British official said it would be cheaper to subsist the unemployed on champagne and venison.
- Although President Lincoln's preliminary Emancipation Proclamation was criticized as inciting servile war and offering a corrupt bargain to the Confederates, a

substantial number of Britons—many of them working men including unemployed cotton workers—publicly approved the final Emancipation Proclamation; and even though most members of the working class lacked the right to vote, the government could not ignore the fact that public opinion generally appeared to favor emancipation.

- In July 1864 Lord Russell, the British foreign minister, instructed Lord Lyons, the British minister in Washington City, to confer with British representatives in Canada about the possibility of a Federal attack against Canada. Lord Russell stated that it was "probable" that General Grant, failing to reach Richmond, would be "obliged to retire," and he apparently worried that the Federals would march on Canada.

International law placed obligations on neutral nations that required their governments to behave in a manner that was, in fact, neutral, but it did not require the neutral nation's citizens to cease their trade with the belligerents. Thus, British merchants were at liberty to purchase Confederate cotton and to sell most kinds of munitions and other goods to the Confederates. Similarly, the Federals possessed various rights under international law to prevent these transactions, which put the Federals in conflict with the neutral nations that sought to protect their nationals and their national interests.

International law viewed the oceans as a vast system of highways that was open to the peaceful use of all nations. In peacetime, those who attacked or seized ships at sea were pirates, common enemies of all civilized nations who could be hunted down by anyone. A state of war altered these circumstances. The warring parties were belligerents, and the other nations with no active participation in the war were neutrals, and all nations had rights and obligations consistent with their respective roles. A belligerent's public ships—those operated by its government—had the right to seek and capture or destroy the enemy's public and private ships wherever found other than within the territorial waters of a neutral nation. By granting letters of marque, a belligerent could convert its private ships into privateers that could lawfully seek out and capture or destroy the enemy's ships.

Neutrals retained a conditional right to trade with belligerents during wartime. The state of war empowered the belligerents to police and interfere with the neutrals' ability to trade with enemy nations. Belligerents could stop neutral ships on the high seas and search them for the enemy's goods (owned by an enemy government or national as opposed to a neutral government or national) or "contraband of war." If the cargo contained either enemy goods or contraband, the ship could be detained, and the enemy's goods and contraband could be condemned as a prize. Generally, only those items were at risk, but certain circumstances might cause the ship to be condemned as well—if, for example, the owner of the contraband also owned the ship or if the ship performed acts (beyond carrying cargo) that aided the enemy.

A key question was what constituted "contraband"? Scholars and jurists formulated various seemingly simple conceptual answers but failed to provide practical clarity. Goods could be viewed as being (a) purely martial, and thus contraband; (b) purely pacific, and thus not contraband; or (c) of a mixed nature, which might be contraband in some circumstances and might not be contraband in others. Inasmuch as the prize courts in the belligerent nation that seized the contraband made the rulings about what constituted contraband, we cannot be surprised that slim distinctions and contradictions abounded. The problem was that the goods that sustain a society render it more able to wage war, so almost all goods can be viewed as having a potentially martial aspect.

A belligerent also could impose a blockade that attempted to cut off all communications and commerce with the enemy. Once the blockade was established, the blockading nation's navy could capture any vessel sailing to or from a blockaded port. The act of attempting to run the blockade made the vessel and all of its cargo subject to condemnation as a prize. In keeping with the British practice, the Federal Navy gave notice of blockade to the enemy port and allowed the neutral vessels two weeks to sail. During that time, neutral ships seeking to enter the port would be stopped, advised of the blockade and turned away, and any neutral ship that persisted in attempting to enter could be taken as a prize. After two weeks the blockaders presumed that knowledge of the blockade was widespread, and all ships attempting to enter or depart became subject to capture.

The existence of a lawful blockade permitted the belligerent to make a prize of any ship and cargo headed to a blockaded port; in the absence of a blockade, only the enemy's goods and contraband were at risk; when a blockade existed, the neutral's ship and her entire cargo became the prize. A lawful blockade was one that was maintained with a naval force sufficient to create a substantial risk of capture. The requirement was intended to prevent paper blockades that were declared but not maintained by a substantial and continuous naval presence.

Early in 1862 Britain declared that the Federal blockade was effective as a matter of international law. Although the Confederates complained that over 100 foreign vessels had come to their ports from November 1861 through January 1862, Britain observed that the blockade was maintained continuously, the blockaders created a risk of capture and no vessels were permitted to pass (other than those permitted by law). A precedent that did not require maximum force or effectiveness, in fact, served British interests. The cotton embargo reinforced Britain's position. The southern press spoke of it openly, but Confederate diplomats were unwilling to concede its existence, so British officials could ask, if the blockade was ineffective, where was the cotton?

As the Civil War began, the Federals possessed a Navy of 42 ships in commission, which was clearly insufficient to enforce a blockade according to the tenets of international law. The government proposed instead to close the ports of the places in rebellion, and Congress enacted a law giving the president the authority to do so. Britain stated that such an act would constitute a paper blockade, and she threatened to force the ports open or recognize the Confederacy, either of which might have led to war. The president acquiesced and resolved to build the Navy as fast as possible. For the first five months of the war, Secretary Welles persisted in asserting that the Federal Navy's operations off the Confederate coast were based upon the law of the United States and not international law, although he acknowledged that such actions would be guided by the principles of international law—the "rules of blockade"—so long as the government found the practice convenient and advantageous. That, for practical purposes, closed the issue.

From that point, Secretary Welles fought within the cabinet to preserve the Federals' belligerent's rights from the British efforts to enlarge neutrals' rights. Britain's threat of war over the *Trent* affair made Secretary Seward amenable to Britain's wishes in maritime matters. When the British complained in August 1862 about the conduct of the blockade, Secretary Seward sent to Secretary Welles a set of instructions—four strongly worded prohibitions intended to preserve neutral rights under the "law of nations" at sea—to be issued to naval officers under the direction of the president. The British probably provided the language to Secretary Seward who obtained the president's approval without consulting Secretary Welles. Secretary Welles revised the pro-neutral rights instructions into strong directions to exercise belligerent rights subject to the established limitations of neutral rights. For example, Secretary Seward's version of one rule said, "That under no circumstances will they seize any foreign vessel within the waters of a friendly nation." Secretary Welles rewrote it as, "That you will exercise constant vigilance to prevent supplies or arms, munitions and contraband of war from being conveyed to the insurgents, but that under no circumstances will you seize any vessel within the waters of a friendly nation."

The British also sought an instruction that public mail bags of a neutral country, even if found on a Confederate vessel or a vessel subject to condemnation, should not be examined or detained but should be sent on to their designated destinations. Federal law, consistent with international law, required masters of prize vessels to "preserve all the papers and writings found on board and transmit the whole of the originals unmutilated to the judge" of the prize court. Nothing in these words or the discussions of international law excepted the mails from the rule. Indeed, the discussions of international law were clear that carrying an enemy's dispatches was a non-neutral act that rendered the vessel liable to be condemned as a prize. The justification to change what appeared to be a settled point of law was that postal services were sending mail on private commercial vessels as opposed to government vessels. Secretary Seward dutifully advised Secretary Welles by letter that he should adopt the proposed rule.

Not long after, a Federal cruiser seized the British steamer *Peterhoff*, bound from Britain for Matamoras just across the Mexican border from Texas, and sent her—with her papers, including a British mailbag—to New York for adjudication. The British government took the position that the mail had been seized in violation of the Federal government policy set forth in Secretary Seward's letter. In the action to condemn the *Peterhoff* as a prize, the Federal district attorney moved to deliver the mailbag to the British government. The Navy Department opposed the government's motion—a remarkable incident of one department of the executive branch of the Federal government opposing another department in Federal court—but the judge ruled that the government had the right to present less than all the evidence in support of its case and ordered the bag released. Having won the point, the British government considered the longer-term implication of its position as a nation likely to mount a blockade and decided, two months later, that only the mail carried on government mail packets, as distinct from ordinary merchant vessels, should be immune from inspection.

The British government also sought to bolster its neutral rights by urging the Federal government to issue instructions to its blockade commanders to restrict certain blockading tactics. As originally proposed, the instruction provided that "You will avoid the reality, and as far as possible the appearance, of using any neutral port to watch neutral vessels and then to dart out and seize them on their departure." Although neutrals generally prohibited the cruisers of two different belligerents from departing the neutral's harbor within 24 hours of each other, no similar rule applied to cruisers and merchant vessels, which meant that a cruiser could trail a suspected blockade runner out of port and detain her once she passed outside of territorial waters. Blockade runners countered this possibility by departing at dusk and remaining in territorial waters until out of the cruiser's sight. Secretary Welles had instructed his commanders repeatedly to respect neutral waters. His instructions were consistent with international law, and he emphasized to those commanders whose squadrons were most likely to visit neutral ports that a capture, made by a cruiser lying in wait within or near neutral waters "might not be recognized as valid." These words represented the shade of difference that the "dart out and seize" instruction sought to erase. During the wars against Revolutionary and Napoleonic France, British cruisers had stood off American ports and captured American vessels. Secretary Welles argued in a memorandum for the president that, by acquiescing to the proposed instruction, the Federal government would be restricting its own operations without obtaining a comparable restriction on the operations of the Confederate cruisers and privateers and, without obtaining a commitment, that a potential belligerent, such as Britain, would be bound in a future war. Moreover, as the British government had argued on many prior occasions, any diplomatic attempt to interfere with a case in prize court that had not fully run its course, including any appeal, was premature. Ultimately, the Navy Department did not issue any version of the "dart out and seize" instruction.

When Secretary Welles received the British proposed "dart out and seize" instruction from Secretary Seward, he went to see the president and found him weary. The president became disturbed when the secretary opened his portfolio to reveal a large quantity of papers. As Secretary Welles explained the matter briefly, the president recalled the instruction vaguely. Intrigued by the matter, the president asked for papers that he could read to understand it more fully. Secretary Welles prepared a memorandum and a letter. The first was written in a manner that suggests that it was intended to be shared with the British government. It set forth a close analysis of the prize court precedents that argued that the logic of the several decisions was to uphold the tactics complained of. Although asserted forcefully, the position was weak and ultimately unpersuasive. The second apparently was intended for the president and his advisers only. As described above, it stated that the Federal government should not act unilaterally to forbid a naval practice that was not inconsistent with international law while leaving the other nations of the world free to continue employing it.

Secretary Welles was less successful in his attempt to detain neutral nationals engaged in blockade running. As with any business enterprise, blockade running depended upon capital, labor, experience and expertise, all of which were provided to a significant degree by neutral nationals. The capture of a blockade-running vessel led to its condemnation and the removal of capital from the enterprise. If the officers and crew of a captured blockade runner could also be detained as prisoners of war, not only would the pool of labor, experience and expertise be reduced, but the increased risk might have deterred others from becoming involved in a blockade-running enterprise. Here, again, the letter of international law gave a protected status to neutral nationals, who were to be detained only in such numbers and for so long as their testimony was required in a prize court. Nevertheless, Secretary Welles tried to maintain a more onerous rule. On January 11, 1864, Secretary Welles issued an instruction that "Henceforth, British blockade violators will not be released, but detained, and any orders which you may have received inconsistent with this are hereby revoked." On May 9, 1864, he issued more detailed instructions with respect to the detention or release of people found on captured blockade runners, differentiating among them based upon whether they were United States nationals or foreign nationals and upon the apparent threat that they posed. The principal elements of these instructions with respect to foreign were as follows:

- Foreign nationals captured in neutral vessels were prisoners of war but were entitled to immediate release unless guilty of belligerent acts. Those needed as witnesses in prize cases were to be held until their testimony was secured.

- Foreign nationals captured in a vessel that hid its nationality or that flew a Confederate flag were prisoners of war if they were officers or crew. Foreign nationals who were passengers without any interest in the vessel or cargo and not connected with the Confederacy were to be released.
- People habitually engaged in violating the blockade, even if not serving as a part of the vessel's crew, were to be detained.

Secretary Welles modified his instructions, apparently in response to international pressure, to say that "The fact of [foreign nationals] having repeatedly violated the blockade gives no authority for their detention." In practice, however, long detentions continued because the crews of the blockade runners lied about the nature of their voyages and were suspected of lying about their nationality.

Toward the end of the war, the courtesies extended pursuant to international law to foreign nationals captured while engaged in blockade running came to an end. By the spring of 1865 the principal Confederate ports were closed, and blockade running was ceasing to be a viable enterprise. In March 1865 President Lincoln issued an order directing that all people who had been "holding intercourse or trade with the insurgents by sea" would be arrested and held as prisoners of war until the end of the war if they were found in the United States more than 12 days after the date of the order if on the Atlantic side and more than 40 days on the Pacific side.

The northern states were themselves a source of supply for the Confederacy, both in the form of smuggling across the long common border shared with the Confederacy and through the hands of merchants of neutral nations. With the start of the Civil War, the Federal government took steps to curtail all commerce and communications with the Confederacy. The Federal Treasury Department assigned customs officers to inland posts at river ports and railroad depots on routes that led into the Confederacy with instructions to prevent the passage of "arms and munitions of war, provisions, and other supplies to persons and parties in those States in open rebellion against the constituted authorities of the Union." As noted previously, some trading through the lines persisted throughout the war and was condoned from time to time by the highest levels of the Federal government.

These restrictions acquired a substantial international impact when the Federal government learned that anthracite coal was being shipped from Federal ports to Confederate ports and to other ports for use by the Confederates. (Anthracite produces little smoke, so a steamship fueled with it was less at risk of being spotted at sea, which was an obvious benefit to blockade runners.) On April 14, 1862, the Treasury Department instructed customs officers to "Clear no vessel with anthracite coal for foreign ports nor for home ports south of Delaware Bay till otherwise instructed." A month later the Federal government modified the instruction to prohibit shipments of coal to points south of the Delaware Bay to the midpoint of the coast of Brazil and east to roughly the middle of the Atlantic Ocean, which embraced the Confederacy, the eastern coast of Mexico, Central America and the northern portion of South America as well as Cuba, Bermuda and the West Indies.

Federal legislation authorized the secretary of the treasury to refuse a clearance to any vessel or other vehicle whatever the ostensible destination if he had "satisfactory reason to believe" any portion of the cargo was intended for the Confederacy. Hiram Barney, the collector of customs for New York, reported that the exports to the Bahamas from New York rose from about $800,000 during the first quarter of 1860 to about $1,200,000 for the first quarter of 1862 and produced evidence that the goods were going to the Confederacy. Mr. Barney noted a number of instances in which vessels or cargoes cleared to Nassau ended up in Confederate ports. He also produced correspondence and papers:

> among which were articles of co-partnership between members of a firm having a mercantile house in London, under the name of Jorss & North, a branch house at Charleston, under the firm of Beach & Root, and an agency or depot at Nassau, were found on the person of one of the members of said firm in London, then on his way to Nassau with instructions how to proceed on his arrival, naming the parties resident there who would be most likely to aid him successfully in transshipping his goods in small vessels, in case their steamers, then on their way, were unable to run the blockade. It was also shown by bills of lading, &c., that about one hundred forty thousand dollars worth of goods had been shipped by their house in London in the British steamers

Memphis and Pacific, in joint account, the proceeds of the sale of said goods at Charleston to be used in the purchase of a return cargo of cotton.

The Federal customs officers began refusing clearances for vessels with certain cargoes for the Bahamas and demanding bonds to assure that the shipments would not go to the Confederacy. Merchants who gave such bonds were often able to repurchase them for a fraction of their face value from corrupt customs officials.

The British government protested these actions, arguing both that the discrimination against British trade violated treaty obligations and that the broad export restrictions were an illegal component of the blockade—international law permitted the Federals to seize contraband wherever they found it, but they could seize other goods only if captured in an attempt to violate the blockade. Thus, the British government argued, the Federal government could restrict the export of contraband, but its attempt to prevent the export of other goods was illegal regardless of their ultimate destination. The Federal government replied that export restrictions did not single out any nation for disparate treatment but, in each instance, were based upon a reasonable factual foundation that the ultimate destination was the Confederacy. Moreover, the treaty obligations of equal treatment were subject to each party's requirements of law, and the trade restriction was based upon the domestic authority of a sovereign nation, not a blockade measure existing under international law. Accordingly, the Federal government was competent to restrict trade with respect to all classes of people and all classes of goods.

The matter remained unresolved, although it did not rest. British merchants who had given bonds to enable them to ship goods to the Bahamas complained when the customs officers in New York declined to cancel them, but eventually the regulations were amended to provide for cancellation. In January 1864 customs officers refused to clear a large shipment of flour, pork and butter to Newfoundland. Canadian ports were used to transship goods to the Confederacy, and, by this time, the Confederates and sympathetic British subjects had begun using Canada as a base for hostile operations against the Federals on land and at sea, so the Federal government was not receptive to British complaints about the refusal.

The Federal customs officials' conclusions were well founded. A report submitted to the British government by Rawson W. Rawson, governor of the Bahamas, shortly after the end of the Civil War reflected the increase in imports and exports of the Bahamas during the war years (Tables 7-1 and 7-2).

TABLE 7-1: IMPORTS TO NASSAU IN £			
Years	United Kingdom	British North America	United States
1860	25,442	762	92,800
1861	51,025	166	136,002
1862	762,627	20,803	352,520
1863	1,054,775	60,797	2,932,945
1864	1,218,914,	51,217	3,772,389

Governor Rawson's report notes that of the 1864 imports, £3,584,587 "represents cotton, imported, with few exceptions, from the two southern Confederate ports of Charleston and Wilmington."

TABLE 7-1: EXPORTS FROM NASSAU IN £			
Years	United Kingdom	British North America	United States
1860	37,901	1,401	79,834
1861	43,901	2,304	104,027
1862	304,733	547,258	134,579
1863	2,124,539	978,681	155,014
1864	3,511,208	889,470	93,314

Proximity of the United States to the Bahamas made them active trading partners under normal circumstances. Governor Rawson's report comments that the "exports to British North America represent goods cleared chiefly for St. John's, New Brunswick, but intended to be run through the blockade."

Both the British national government and British colonial authorities used coal to influence the operations of Federal and Confederate cruisers, according them differing treatment. In December 1861 the Federal Navy sent two ships loaded with coal to Nassau to establish a depot, but local authorities refused permission to land the coal except on the condition that the coal "should not be reshipped or transshipped on board any United States war vessel during the continuance of the struggles now going on in America." One of the ships was leaking badly, and, when the United States consul asked permission to transfer some of the coal to the USS *Flambeau,* then in port, the colonial authorities "courteously refused," stating that providing coal would be an infraction of Britain's neutrality and noting that Nassau was a short distance from Key West where all the *Flambeau*'s needs could be supplied. They asserted that the sole purpose of the *Flambeau*'s coaling in Nassau was to permit her to blockade the port, which they refused to assist without the express direction from the British government. In the meantime, the authorities permitted blockade runners flying Confederate colors to take on coal. A similar incident occurred in Bermuda.

In January 1862 the British Foreign Office issued further instructions that were intended to prohibit the use of the ports and waters under British jurisdiction "in aid of the warlike purposes of either belligerent." Warships and privateers were barred from Nassau and the Bahama islands unless permitted to enter by special leave of the lieutenant governor or "in case of stress of weather." The instructions prohibited them from making use of the ports and waters under British jurisdiction as stations or places of resort or for the purpose of obtaining warlike equipment, and required them to depart within 24 hours after entry except due to weather or obtaining provisions or repairs. The instructions prohibited a belligerent's vessel from sailing out of British waters until at least 24 hours after a vessel of the other belligerent had departed, and prohibited a warship or privateer from taking on more coal than was sufficient to carry her to the nearest port in her own country or some nearer destination and coal would not be supplied to any warship or privateer until three months from the last time coal was supplied in British waters. Notwithstanding such instructions, Federal naval officers and consular officers reported discourteous treatment and open hostility in British waters and apparent disparities in treatment as compared to Confederate vessels. When the USS *Dacotah* put into Nassau for coal in

September 1862 the governor granted permission on the condition that the Dacotah cruise at least 5 miles from any island of the Bahamas for 10 days after leaving port even though neutral waters extended only 3 miles.

Other nations did not impose similar restrictions: the Federal Navy obtained coal and provisions in the Spanish possessions Cuba and Puerto Rico and in Haiti, an independent nation recently recognized by the United States; and the Navy maintained coal depots at the Dutch possession Curaçao, the French possession Guadeloupe and the Danish possession St. Thomas.

After the Civil War, the United States sought monetary damages from Britain for injuries sustained by its Merchant Marine from the Confederate cruisers—an arbitration referred to collectively as the Alabama claims. The principal complaint was based upon Britain's failure to maintain neutrality as evidenced by British complicity, or the lack of diligence, in permitting Confederate cruisers, built in Britain, to get to sea and attack American merchant vessels. The arbitration panel awarded the United States $10 million in damages (about $185,616,900 in 2019 dollars). The fact that British colonial authorities provided coal freely to the Confederates, but denied it to the Federals, turned out to be a major element of proof sustaining the award.

In addition to attempting to influence Federal blockading strategy and practice generally, the British and, to a lesser extent, the French and, to an even lesser extent, the Spanish, registered objections to the interference of Federal blockaders with the activities of their nations' commercial vessels operating off the Confederate coast—alleging violations of international law and the application of unnecessary brutality in the exercise of belligerent rights of visiting neutral vessels on the high seas and the capture of suspected blockade runners. The complaints generally painted a picture of the neutral's innocence and averred that the facts of the matter were such that the intrusion on neutral rights was an obvious violation by the Federal belligerent. The Federal responses generally were assurances that the facts were clearly otherwise and averred that the actions of the Federal vessels and crews were clearly within rights established by international law. Obviously, all the parties were motivated to shade truth to uphold their interests.

Individual commanders of the naval vessels of neutral nations held personal prejudices about the Civil War that led a few of them to undertake acts that were not strictly neutral. Rumor circulated during the first year of the war, for example, that the HBMS *Bull Dog* had conveyed Confederate naval officers to ports outside the Confederacy where they took command of cruisers that preyed upon Federal commercial vessels. On the other hand, the *Bull Dog* was one of a handful of British warships that seized the *Oreto* (later commissioned as the cruiser CSS *Florida*) at Nassau for violation of British neutrality laws about the same time that the *Bull Dog* was supposed to be ferrying the Confederates. A British court dismissed the case concerning the *Oreto* and released the vessel because she left Britain unarmed and remained that way in Nassau. Later, while the *Oreto* was in the outer anchorage at Nassau, a Federal cruiser came in to investigate her, but the HBMS

Petrel intercepted the Federal cruiser and instructed her to go either into the harbor or out beyond the marine limits. Shortly after, the *Oreto* sailed to a more remote location, again with the assistance of the *Petrel*, where she received her armament and her Confederate commission.

James Magee, the acting British consul in Mobile, enlisted the HBMS *Vesuvius* to transport specie (gold and silver coins) out of Mobile for shipment to Britain for paying interest on Alabama state bonds held by British investors. Lord Lyons, the British minister to the United States, sent instructions to prevent the shipment, but the *Vesuvius* sailed before they arrived. Lord Lyons relieved Mr. Magee of his duties. The event prompted British Admiral Alexander Milne, commanding the North American station, to issue an order reminding his commanders of their obligations as neutrals and noting that a neutral's warships, when permitted by the blockaders to pass into a blockaded port, did not possess any right to convey property through the blockade. The actions by Lord Lyons and Admiral Milne in attempting to stop the shipment, disclosing it promptly to the Federal government and issuing firm instructions to the British squadron, prevented the incident from turning uglier.

In a separate incident, the *Petrel* obtained permission to pass through the blockade and enter Charleston Harbor, where she remained for a month. During that period, a pair of Confederate ironclads, the CSS *Chicora* and the CSS *Palmetto State*, came out of the harbor and attacked some of the Federal ships that were maintaining the blockade. The Confederates claimed to have lifted the blockade, and the *Petrel*, with the British consul on board, confirmed the claim.

Secretary Seward asked Lord Lyons to order the *Petrel* out of Charleston Harbor. When this was done, Admiral Milne assigned the *Petrel* and her commander to duties that were distant from the blockade.

Excess zeal in command was not solely a British trait. Secretary Welles named Commodore (later Acting Rear Admiral) Charles Wilkes as the first commander of the West India Squadron. Admiral Wilkes was an aggressive officer whose involvement in the *Trent* affair gave authorities of neutral governments cause to be wary of him. Secretary Welles advised Admiral Wilkes to avoid visiting British colonial ports whenever possible, but the admiral became the subject of complaints from Britain, Denmark, Mexico and Spain. In addition to making himself a liability, Admiral Wilkes failed to capture any Confederate cruiser and interfered with the Navy's efforts to do so, so Secretary Welles relieved him of command.

The profit opportunity created by the Civil War gave the blockade runners and those who did business with them a strong incentive to favor the Confederates or, more precisely, to favor the continuation of the war that created the profit opportunity in blockade running. Among the foreign public vessels that monitored the Federal blockade of Confederate ports, some might have been persuaded by the economic incentive that motivated their countrymen to lean toward the Confederacy, and some might have been persuaded by their professionalism to maintain an attitude of genuine neutrality, whether aloof or courteous to the Federals. The fact that neutral possessions off the American coast were destinations for blockade runners also gave rise to interests and prejudices coming into conflict. For example, on May 30, 1863, the USS *Rhode Island* was cruising about 20 miles northeast of Eleuthera Island, in the Bahamas, when she sighted a steamer about 8 miles distant and gave chase. As the distance shortened the *Rhode Island* showed her colors and fired a blank cartridge, an instruction to heave to. When the steamer did not show colors or stop, the *Rhode Island* fired a shot ahead of the steamer with her longest-range guns. As Eleuthera came into view, the steamer headed for shore with the *Rhode Island* closing and still firing. The *Rhode Island* reported that she ceased firing when the steamer was about 3.5 miles from shore. The steamer started making dense smoke, and she beached herself.

The steamer was the *Margaret and Jessie* bound from Charleston with a cargo of cotton. The Confederates' version was that the *Rhode Island* started firing live rounds without showing colors, that she pursued the *Margaret and Jessie* into British territorial waters and, when the *Margaret and Jessie* had beached, continued to fire rounds into her from a distance of about 500 yards. They also obtained affidavits from residents of Eleuthera saying that shells fired by the *Rhode Island* struck the island. The British government submitted a copy of the protest made by the master and crew of the *Margaret and Jessie*. Neither side had the monopoly of truth, but the Federal account sounds more credible—if the *Rhode Island* had fired on the *Margaret and Jessie* at a range of 500 yards, she likely would have been destroyed. The protest acknowledged that the wreckers discharged the cargo and succeeded in floating the vessel.

A Federal Navy court of inquiry concluded that the *Rhode Island* did not violate the territorial jurisdiction of Britain. The British government pronounced itself satisfied with the court's conclusion, but the matter did not rest there. The British government noted that the range of a Parrott rifle was 5 miles, which meant that a shot fired from outside the 3-mile limit could endanger life and property on the neutral shore. The British government invited the Federal government to "concur with them in opinion that vessels should not fire towards a neutral shore at a less distance than that which would insure shot not falling in neutral waters or on a neutral territory"—in essence, a request that the Federal Navy cease firing at blockade runners when they were within 8 miles of a neutral shore. Secretary Seward replied that the Federal government was unwilling to renounce unilaterally any advantages of a belligerent, stating instead that any new construction of international law should be reciprocal and binding upon the principal maritime powers. Secretary Seward's response, made in mid-September 1864, is the same that Secretary Welles used a year earlier to persuade the president not to acquiesce in the "dart out and seize" rule proposed by the British.

The neutral powers insisted upon the observance by the Federal blockaders of the customary forms of notice. Accordingly, Secretary Welles instructed one of his squadron commanders that they must fire "shotted guns" (that is, guns loaded with solid shot) across the bows of vessels approaching the blockaded coast and sometimes upon the high seas. The insistence on such a warning must have been galling inasmuch as the blockade was supposed to have been notorious within weeks after it was imposed. In giving this instruction, however, Secretary Welles remained in touch with the practical realities of the situation when he added, "A vessel approaching the blockade at night without making the usual signals can not be regarded as entitled to the customary courtesies of the sea."

Even though Confederate cotton shipped out to foreign destinations, and munitions and other supplies shipped in, the trade through Matamoras, Mexico, was neither covert nor blockade running since Matamoras was a neutral destination. Ordinary freighters, rather than purpose-built blockade-running vessels, carried the trade. Federal cruisers could stop and search neutral vessels sailing to the Mexico side of the border while in international waters, and contraband or goods intended for the Confederates could become a prize, but the ships themselves were not at risk. Once in Mexican waters, neutral vessels were beyond Federal reach. Cotton owned by the Confederates and shipped from Mexico was at risk of capture once at sea, but the vessel carrying the cargo was not, although it could be sent to a Federal port and detained until a prize court adjudicated the capture. On the other hand, cotton owned by neutrals and shipped from Mexico was not subject to capture.

In the fall of 1862 Federal officials became aware of communications in the British shipping community about the Confederates' ability to receive goods in Mexico and to pay for those goods in cotton. Flag Officer David G. Farragut, who took command of the West Gulf Blockading Squadron in February 1862, approved of aggressive action toward the shipping through Mexico. He wrote to Secretary Welles, "you will perceive that the English and French are not complaining of the laxity of the blockade, but the rigidity of it." The blockaders captured ships bound to and from Mexico and sent them north, but so long as the shippers exercised caution and observed the legalities, the prize courts declined to condemn their cargoes. Despite the harassment, traffic through Mexico remained active. The spring of 1863 found between 180 and 200 vessels at anchor off the mouth of the Rio Grande, whereas before the war, a half dozen vessels might have arrived during a whole year. The vessels represented all nations, including American vessels that cleared from northern ports. Secretary Welles asked Secretary of the Treasury Salmon P. Chase whether regulations might be drawn to interfere with the traffic in Texas cotton through Mexico. In May 1863 Lord Lyons warned Secretary Seward that the Federals appeared to be "systematically endeavouring by fair means and by foul to stop our [British] trade with Matamoros." Although the Federals had given assurances of neutrals' rights, and the Navy Department had given appropriate instructions, Lord Lyons stated that "those instructions were apparently set at nought by the U.S. officers." Lord Lyons wanted the Federal government "to make the subordinate officers feel the effects of the displeasure of the Government, when they violated neutral rights." The continued British naval presence near Matamoros after 1863 ended the Federal seizure of vessels there.

With the fall of Vicksburg in July 1863, Federal forces commanded the entire length of the Mississippi River, cutting the Confederacy in two and interrupting communications between the two halves. The eastern portion of the Confederacy lost the supply of western beef and horses, and the western portion lost the supply of arms, but the Confederate Ordnance Bureau had arranged for contracts to supply arms to Texas or by way of Mexico directly from Europe. Transporting cotton from Texas to Mexico—especially after the Federals captured Brownsville, Texas, in November 1863 and employed land forces in an attempt to suppress the cross-border trade—increased its cost at Matamoras (36 cents a pound) compared to its cost in Confederate ports (about 7 cents a pound).

The frustration of the Federal officers assigned to blockade duty off the Rio Grande was evident in a nonofficial letter written by Captain Henry Rolando of the USS *Seminole:*

> I could not stand the hocus-pocus of those English men there; they were in open daylight landing cargoes of contraband, with the connivance of the Mexicans, and sending into Texas everything they could which would aid the rebellion....
>
> No one knows, who has not been to the Rio Grande, the immense value that river is to the Southern Confederacy. Steamers are constantly landing cargoes—contraband—which go to Texas in the small steamers and barges, which are towed up the river and landed; on returning, bringing cargoes of Confederate cotton—just as bad as contraband, as it buys the arms and munitions of war....
>
> It galled me to have a scoundrel of an Englishman crying out "neutrality," when he knew as well as I did that he was trafficking in the blood of my countrymen and making money out of our misfortunes...

> When it is so palpable to any one of our statesmen that Mexicans are hourly, through the custom-house at the Rio Grande, violating their neutrality by assisting and giving aid and comfort to the enemy, I think it devilish hard that any plea for the observance of a neutrality, which they don't observe themselves, should be listened to by our Government, and I hope Mr. Seward will remind them of what they are now doing, and have done in the way of violating neutrality.

Captain Rolando had captured the *Sir William Peel* near the mouth of the Rio Grande. He claimed the capture had been made in American waters based upon bearings taken at the time of the capture. The Federal Supreme Court eventually found that the *Sir William Peel* had been in Mexican waters and held that her cargo—nearly 1,000 bales of cotton—was owned by neutrals and thus could not be condemned.

Captain Rolando's bad luck did not end there. The mail bag containing his letter was given to a blockade runner who represented himself as a Federal merchantman.

The Confederates sent the letter to London for publication, with one copy to the British admiral commanding in the Gulf and another to the Mexican authorities at the Rio Grande.

Military channels also conveyed diplomatic inquiries. British Admiral Milne wrote to an officer in command of the West Gulf Blockading Squadron asking whether Federal cruisers had been prohibited:

> From capturing ships for having carried contraband of war after the contraband has been actually landed ... From capturing ships lying in Mexican waters; and ... From claiming and exercising the right to seize neutral ships lying within three leagues of the coast of Texas for alleged trading with the enemy, irrespective of any questions of blockade or of contraband.

(The mention of "three leagues" is a reference to a provision in the Treaty of Guadalupe Hidalgo in which the United States and Mexico set the extent of their respective territorial waters. The Federals maintained that for purposes of the blockade 1 marine league—3 miles—was the relevant distance.) As phrased, Admiral Milne's question presupposed that any taint from a neutral vessel carrying contraband of war was washed away when the contraband was landed. Admiral Farragut responded that the Federals claimed the right to capture any ship that had violated the blockade, unloaded contraband in a blockaded port or was on a return voyage from the blockaded port. Such a vessel remained at risk of being seized until it made its next innocent voyage. The admiral also acknowledged that the Federals did not claim the right to seize neutral vessels in Mexican or other neutral waters.

Canada, although more distant from the Confederacy than Bermuda or the Bahamas, not only served as a neutral base for blockade running but also provided a base from which the

Confederates and their sympathizers attacked the Federals, such as the plot to seize a Federal ship on Lake Erie and use it to liberate Confederate prisoners on Johnson's Island, Ohio, and the raid on St. Albans, Vermont. These actions threatened war between Britain and the Federals and raised British fears that the Federals would invade and conquer Canada.

In November 1863 a group of adventurers—apparently British subjects with pro-Confederate leanings and possessing a Confederate letter of marque—sailed from Canada to New York where they took passage on the steamship *Chesapeake*, a coastwise steamer bound for Portland, Maine. On December 7, while the *Chesapeake* was off Cape Cod, the adventurers seized control of her and headed for Nova Scotia where they proposed to sell her cargo, purchase guns and run them into the Confederacy. They also planned to use the *Chesapeake* as a privateer to prey upon Federal shipping. On December 9, the Federal Navy Department learned of the seizure, and Secretary Welles began telegraphing orders for vessels to pursue. On December 10, Secretary Welles advised that the *Chesapeake* was in Pubnico Harbor in Nova Scotia.

The first Federal vessel on the scene located the *Chesapeake* in the inshore waters of Nova Scotia and took possession of her, capturing three of the adventurers. The commander of the next Federal vessel to arrive had a greater awareness of international sensitivities and persuaded the first commander that the *Chesapeake* should be surrendered to authorities in Nova Scotia. Three other Federal vessels arrived, and all five accompanied the *Chesapeake* to Halifax, where they steamed into the inner harbor and turned her over to British authorities. Two more Federal warships had joined the flotilla along the way. A dispute arose when the Federal officers balked at surrendering the three prisoners, and, at one point, the local British authorities threatened to fire upon the Federal vessels with their shore batteries. The fire of seven Federal gunboats would have devastated Halifax. When the Federals eventually surrendered their prisoners, the hostile, pro-Confederate crowd permitted one to escape, and a Canadian court released the others. Diplomats soft-pedaled the events. Secretary Seward, for example, assured the British government that the Federal invasion of Canada had not been instigated by the Federal government, which must have taken exceptional diplomatic finesse in light of Secretary Welles' hot-pursuit telegrams.

8. The Campaign: 1861-1863

Stories of battles and campaigns dominate published military histories for the obvious reasons that they are intrinsically dramatic and significant events, even when they are indecisive. For the same reasons, the stories of battles on the Confederate coast constitute a large part of the literature, even though they represent only a small part of the overall history. Nonetheless, they bear revisiting here, even in brief, because understanding what occurred in them is essential to comprehending the larger struggle over control of the Confederate coast. Such battles generally consisted of combined operations by Federal Army and Navy forces against fixed Confederate defenses on the coast and the occasional fight between Federal and Confederate warships. In the combined operations, the Federal Navy generally brought enormous firepower to bear, and the Federal Army, operating under the cover of Navy fire, was generally able to secure objectives on land. Exceptions arose when the Confederates had at their disposal sufficient resources to mount an attack or sustain a persistent defense, and also when the tactical situation gave the Confederates an offsetting advantage.

These high-profile events occurred within the context of the blockading and blockade running previously described, and it is the effect on blockading and blockade running that gave these events their significance in the war. The persistence and volume of blockade running is the measure of the effectiveness of the Federal operations along the coast. The various elements of the conflict that we have reviewed up to this point combine to reveal the larger pattern of events. In evaluating the port seizures in the context of blockade running, the following discussion looks first at the five-month period from the start of the war in April through August 1861 and thereafter in four-month segments from September 1861 until the end of the war. Each trip is a successful or failed effort by a steamship to make a voyage through the blockade to or from a Confederate port. The destinations were grouped around the principal Confederate ports and their environs: Wilmington, Charleston, Savannah, St. Marks, Mobile, New Orleans and Galveston. The "Losses" category includes blockade runners captured or destroyed by blockaders patrolling the coast as well as other losses and captures at sea.

TABLE 8-1: APRIL-AUGUST 1861	Attempts	Successes	Losses
All Confederate Ports	0	0	0
Timeline: The war began with the bombardment of Fort Sumter on April 12, 1861. The first principal battle, Bull Run on July 21, 1861, showed that the war was likely to be long.			
Note: The statistical information in Tables 8-1 to 8-8 is derived from S.R. Wise, *Lifeline of the Confederacy* at 233-284.			

1861. Although steamships did not begin to challenge the blockade until September 1861, the Federal blockaders began making captures of other types of vessels along the Atlantic Coast in May and along the Gulf Coast in June (Table 8-1). The coastwise trade continued through this period, and some of the coasting vessels were steamers. The Federals had established at least nominal blockades of the principal Confederate ports by late July.

Early in the war, the Confederates attempted to defend their entire perimeter, although aside from the principal ports, most places were only lightly defended. The initial Federal attacks on the Confederate coast were at the periphery rather than at places of strength so that the naval resources that the Federals possessed, even early in the war, overwhelmed the Confederate defenders. The Federals directed their first principal coastal attack in August 1861 against Hatteras Inlet, an entrance to the Pamlico Sound in North Carolina, a refuge for Confederate privateers preying upon Federal merchant vessels. The Confederates defended the inlet with two sand forts. The Federal expedition included both Army and Navy forces, but wind and surf prevented the troops from going ashore. Navy guns drove the Confederates from one fort and forced the surrender of the other.

General Scott's Anaconda Plan to split, surround and starve the Confederacy into submission focused on the Mississippi River as the principal avenue for advancing military actions into Confederate territory. The river network provided an economical means for the Federals to advance deep into Confederate territory. The superiority of Federal resources meant that the Federal forces would have better and more numerous boats and weapons with which to maintain their control of the portions of the river network under their command and to push for even larger areas of control. Unless the Confederate forces could assert a local control of the river network—a feat that would be more difficult for them owing to the difficulty of moving their forces overland—the river network represented a supply route that could not be interrupted. The public, and hence political, demand for immediate actions soon led to the initiation of Federal advances on Confederate forces at various points around the perimeter of the Confederate states. The use of the river network and the drive down the Mississippi River remained an important element of Federal operations.

Although Federal strategic thinking and operations were largely directed at pushing through the river network and down the Mississippi River, attention also was given to pushing up the Mississippi River through New Orleans. As a part of the initial blockading activities that Federal forces undertook in the Gulf of Mexico, on May 26, 1861, they served notice of the establishment of the blockade on one of the mouths through which the Mississippi River flowed into the Gulf. On September 19, 1861, Federal boats conducted a reconnaissance of the Head of Passes, the point below New Orleans at which the Mississippi River divided into three principal streams and flowed into the Gulf of Mexico. Federal forces occupied the Head of Passes on October 2, 1861. This occupation was not uncontested. A Confederate steamer attacked Federal forces there on October 9, and an attack on October 12 drove Federal forces from the Head of Passes temporarily. Although the occupation by Federal forces of the Head of Passes closed the Mississippi River to most traffic below New Orleans, New Orleans remained active as a port of entry for shipments to and from the Confederacy through other water routes to the Gulf and remained useful due to the availability of railroads to carry goods to and from other points in the Confederacy.

One alternative route to New Orleans was through the Grand Island Pass that connected the Mississippi Sound with Lake Borgne to the east of New Orleans. The terminus of the Mexican Gulf Railroad lay at Proctorsville on the lake's southern shore, and the railroad ran to a levee of the Mississippi River about 12 miles from New Orleans. Lake Borgne also connected with Lake Pontchartrain by a pair of crooked channels called the Rigolets. Lake Pontchartrain lay due north of New Orleans and its southern shore was connected to the city by two railroads, a canal and a bayou.

The occupation of the Head of Passes did not mean the blockade of the river was complete. About 7 AM on February 19, 1862, with the Mississippi River thick with fog, the steamer *Magnolia* descended the river and was observed by the USS *Brooklyn* and the USS *South Carolina* only because her masts and smokestack extended above the fog. The Federal vessels gave chase and captured the *Magnolia* in the afternoon. The pursuit by both Federal vessels left the Head of Passes unoccupied and possibly permitted other blockade runners to escape. In explaining his regret for the *Magnolia*'s escape from the river, Flag Officer McKean noted that the Confederates had dredged channels that permitted vessels that drew 6 or 7 feet of water to escape New Orleans into the Gulf of Mexico without descending the length of the Mississippi River. The Magnolia drew 8 feet, which compelled her to descend the river.

Another Federal Army-Navy expedition attacked and captured Port Royal Sound in November 1861, which the Federals used as a port of refuge for the South Atlantic Blockading Squadron. A storm scattered the vessels of the expedition while on their way to Port Royal, and most of the Army's landing craft were lost, so the burden of the attack fell solely upon the Navy. Two Confederate forts on the inner shores of the sound and three gunboats defended against the attack by a flotilla of 14 Federal gunboats. A squadron of five Federal gunboats entered the sound and took a position where they could fire into one of the forts and keep the Confederate gunboats at bay. Another squadron of nine Federal gunboats steamed in a circle, firing first at one Confederate fort and then at the other. During the course of a day and a night, the Confederates abandoned both forts.

About this time, the Confederates abandoned their perimeter coastal defense, generally moving their forces inland and concentrating their resources only at key points on the coast.

Both the Federals and the Confederates had taken an interest in Ship Island since the early days of the war. Ship Island, off the coast of Mississippi, offered the only deepwater port between the Mississippi River and Mobile Bay. On July 4, 1861, the USS *Massachusetts* passed Ship Island without noticing anything in particular and returned on the evening of July 8 to discover that the light in the Ship Island lighthouse had been extinguished and that Ship Island had been occupied by Confederate troops. The *Massachusetts* exchanged cannon fire with the Confederates on Ship Island on July 9 and then proceeded to nearby Chandeleur Island where the crew removed the lens and lighting apparatus from the lighthouse to prevent them from falling into Confederate hands.

On August 9, 1861, just a month later, the Federal Blockading Conference delivered its first report on the Gulf Coast, which discussed in detail the approaches to New Orleans and posed recommendations for proceeding with the blockade. The first recommendation that the conference made in this report was the "complete military possession" of Ship Island, which, the conference stated, was "the key to the blockade and possession of Mississippi Sound and the control of the coasts of Mississippi and Alabama."

On the evening of September 16, 1861, the Confederates burned their barracks and destroyed the lighthouse, and, on September 17, they evacuated Ship Island. Lieutenant Colonel H.W. Allen, the Confederate commander of Ship Island, left behind the following letter addressed to the commander of the *Massachusetts*:

> By order of my Government, I have this day evacuated Ship Island. This my brave soldiers under my command do with much reluctance and regret. For three long months your good ship has been our constant companion. We have not exactly "lived and loved together," but we have been intimately acquainted, have exchanged cards on the 9th day of July last. In leaving you to-day we beg you to accept our best wishes for your health and happiness while sojourning on this pleasant, hospitable shore. That we may have another exchange of courtesies before the war closes, and that we may meet face to face in closer quarters, is the urgent prayer of very truly, your obedient servant.

Federal troops arrived at Ship Island on December 3, 1861, and spent the next several days landing. Ship Island became a staging area for the Federal move up the Mississippi River against New Orleans and a Federal port of refuge for ships blockading the Mississippi and Alabama coasts.

In November 1861 the Federals occupied Tybee Island at the mouth of the Savannah River, which the Confederates also had abandoned.

TABLE 8-2: SEPTEMBER–DECEMBER 1861	Attempts	Successes	Losses
North Carolina (Wilmington)	1	1	0
South Carolina (Charleston)	5	5	0
Georgia (Savannah) and East Florida	3	3	0
West Florida (St. Marks)	0	0	0
Alabama (Mobile)	0	0	0
Louisiana (New Orleans)	3	3	0
Texas (Galveston)	1	1	0
Lost, Destroyed or Captured at Sea	0	-	0
Totals	13	13	0

Blockade running by steamships began in the period September through December 1861 (see Table 8-2) with a total of 13 voyages, all of them successful, and most of them involving Charleston (5), New Orleans (3) and Savannah (3)—roughly one voyage in or out of the Confederacy every week. The Federals served formal notice of their blockade on the Confederate ports, but they had difficulty in maintaining a continuous blockading presence.

1862. In March 1862 the Federals occupied Amelia Island and Fernandina in Florida, places that the Confederates abandoned. The Federals also took Jacksonville, but they soon abandoned it because it was too exposed and their forces there were threatened by nearby Confederates.

In early 1862 the Federals undertook two major coastal operations that built upon the earlier capture of Hatteras Inlet and the occupation of Tybee Island, respectively. A joint Army-Navy expedition under the command of General Ambrose E. Burnside put Pamlico and Albemarle Sounds in North Carolina under Federal control. The 12,000 Federal troops were more than all the Confederate forces defending North Carolina, and the 80 vessels composing the naval portion of the expedition—gunboats, transports and auxiliary vessels—possessed more firepower than the Confederates could return. The Federals captured Roanoke Island in February and New Berne in March. In April they captured Beaufort, one of two deepwater ports in North Carolina. The Federal Navy used Beaufort as a base and coaling station for the North Atlantic Blockading Squadron. The Federals also mounted a siege operation against nearby Fort Macon, the masonry fort that defended Topsail Inlet, one of two deepwater inlets on the North Carolina sounds. The Federals began their barrage on April 25, 1862, and the fort surrendered the next day.

The Federals also began operations against Fort Pulaski, the masonry fort that defended the principal water access to Savannah. Much of the area surrounding Fort Pulaski was marshy and therefore unsuited for constructing batteries. The possession of Tybee Island was key to the operation because it provided dry land from which the Federals' rifled cannon could reach the fort, but beyond the reach of at least some of the fort's guns. In preparation for the attack, the Federals secured control of the region upriver to prevent Confederate reinforcements from reaching the fort. Once the fort was cut off, the Federals erected their principal batteries. The Federal bombardment began on April 10, 1862, and the Confederates surrendered Fort Pulaski the next day at noon.

The capture of Forts Macon and Pulaski showed the science of siegecraft brought up to date in 1862. The accepted military wisdom when the forts were built was that a properly conducted siege generally resulted in the capture of a fort, and, as noted, the national defense plan expected that the defenders would hold out for about two weeks. In the 1781 siege of Yorktown that ended the fighting in the American Revolution, digging the first parallel began on October 6, and the artillery barrage began on October 9 with fighting and digging continuing until the capitulation on October 19. By 1862 the rifled cannon firing exploding shells from a great distance could breach the walls of the masonry and stone forts. In the case of each capture, a large amount of time was expended in preparing for the attack, but the bombardment itself lasted about one day and rendered the fort defenseless.

New Berne was not a major Confederate port, but, had it not been captured, it might have grown into one, as did Wilmington. Savannah was a major Confederate port, but when the Federals occupied Tybee Island, they cut off the main access to the port, and the subsequent capture of Fort Pulaski bolted the door shut. Of more importance were the contests for control of the principal seaports that were essential to the long-term survival of the Confederate nation—New Orleans, Charleston, Mobile and, eventually, Wilmington.

New Orleans, located on the bank of the Mississippi River, sat about 100 miles from where the river flowed into the Gulf of Mexico. In 1862 about 10 different routes gave access by water from the Gulf to New Orleans. In the autumn of 1861—after the expedition to North Carolina had captured Hatteras Inlet and while the expedition against Port Royal was being organized—the Federal government determined to move against New Orleans. The Federal Navy acquired a flotilla of schooners and armed them with 13-inch mortars and some smaller guns. Secretary Welles assigned Commander David D. Porter to command this flotilla, and he appointed Captain David G. Farragut, newly elevated to the grade of flag officer, to command the West Gulf Blockading Squadron with instructions to capture New Orleans.

Forts Jackson and St. Philip, which defended the Mississippi River access to New Orleans, occupied opposite banks of the river about 70 miles below the city. Fort Jackson was a five-sided masonry fort with bastions at the corners. The fort was surrounded by a wet ditch, and its scarp wall rose a little over 14.5 feet above the surface of the water in the ditch. The parapets were not carried around the flanks of the fort or the faces of the bastions, which limited the space for musketry. The two walls of the fort that faced the river were constructed of casements that could accommodate a total of 16 guns. Fort Jackson and its outer works could accommodate 127 heavy guns, of which 111 guns could fire on the river. Aside from the 16 guns in casements, all the guns were mounted en barbette. Fort St. Philip was an irregular four-sided work, about 150 by 100 yards. Together with its external earthen batteries, Fort St. Philip could mount a total of 72 guns that fired on the river, all mounted en barbette. At the time they came into Confederate possession, the forts mounted 52 guns, a mix of 32-pounder and 24-pounder smoothbores. A Federal general familiar with the forts and the guns available to the Confederates said he expected the armament would contain few guns over 32-pounders. Such weapons did not pose a significant threat to warships unless they fired hot shot, which both forts were able to do. "Still," he observed, "it is not a trifling undertaking to pass so large a number of guns at such close quarters."

The forts commanded about 3.5 miles of the river, and the general estimated that a squadron would need about 30 minutes to cover that distance. He concluded that daylight passage was too hazardous and proposed the attempt be made at night. Ships firing 9-inch and 11-inch guns firing case and canister should make the barbette guns, and even the casement guns, "untenable." The presence of an obstruction in the river could make passage by the forts "almost impracticable," although maintaining an obstruction would be difficult when the river was running high, as it did each year starting in March. The great volume of water, carrying solid objects like fallen trees, would exert a strong pressure on any barrier intended to prevent passage by vessels upriver.

Flag Officer Farragut's squadron consisted of 17 steam-powered warships that mounted a total of 192 guns. The mortar flotilla consisted of 19 mortar vessels and seven steamers. The Federals also had 15,000 troops on Ship Island awaiting deployment to New Orleans. Flag Officer Farragut instructed his commanders to remove the uppermost masts and rigging, leaving only enough to suspend the lower sails; mount guns fore and aft where the broadside batteries did not fire; place "grapnels" (grappling hooks) in boats to tow off fireships; trim the vessels lower at the bow so they would remain headed upstream if they touched bottom; prepare pumps and hoses and drill the men to extinguish fires; and mount boat howitzers in the foretop and the maintop on boat carriages for firing abeam.

Upriver, the defenses of New Orleans included a ring of earthworks. The Confederates also had 13 vessels, all of them civilian boats converted to military use, mounting a total of 24 guns, but these were divided into three separate commands. The two Confederate ironclads at New Orleans remained incomplete. The *Louisiana* was afloat, and her iron armor had been bolted on, but she could not propel herself. She was moored at the riverbank just north of Fort St. Philip where workmen mounted her 16 guns. Altogether, the Confederate forts, batteries and vessels mounted 166 guns.

The Confederates obstructed the river with a raft composed of large tree trunks held side-by-side by heavy chains anchored to the riverbank. The substantial amount of floating debris borne by the increased seasonal flow of the river broke the raft and washed most of it away in March. The Confederates replaced it with a raft built of schooners and chains.

The Federal vessels began crossing the bar off the Mississippi River by mid-March 1862. Surveyors established firing stations for the mortar vessels, just down the river and around a bend from the forts. The Confederates had cleared the area immediately around the forts of trees, but a thick forest concealed the firing stations. Although the mortar crews could not see their targets, they knew the direction and the distances to them. The Confederates sent soldiers to fire on the mortar vessels, but the adjacent land was flooded, making passage difficult, and the fire of Federal gunboats drove the Confederates away.

The bombardment began on the morning of April 18, 1862. The forts returned the fire, and, although they fired blind, and only the forwardmost mortar vessels were within range, the Confederates sank one mortar vessel. The mortars ceased firing at darkness the first day, but their firing continued for the next six days and nights. Federal gunboats broke the chains that held the raft together. Only a reduced rate of firing and a deserter's report of the damage at Fort Jackson indicated that the forts were weakening. With mortar ammunition running low, Flag Officer Farragut decided to move upriver.

The Federal squadron advanced at three in the morning on April 24 in two parallel lines, firing case and canister into the forts as it passed. The night was dark, and, after the firing started, gun smoke covered the river, so all the gunners aimed at the flashes of the enemy's weapons. Above the forts, the Federals encountered the Confederate flotilla. Two of the Confederate vessels rammed the USS *Varuna,* which ran ashore before sinking. The USS *Hartford* ran aground, and a fire raft ignited her side and rigging. The crew extinguished the fire and backed her into deeper water. The Confederates lost three vessels in the fight, and their remaining vessels retreated up the river.

Two field artillery emplacements fired on the Federal vessels as they passed, but the guns were too few and too light to stop them. The Federal squadron steamed up to New Orleans and demanded its surrender. Passing above the forts without reducing them was a calculated risk. Unless supplies could be brought to the Federal vessels, the only way to preserve their effectiveness was to pass back down through the gauntlet, placing the vessels at risk again and giving up what had been obtained by running past the forts in the first place. Fortuitously, the Confederate garrison of Fort Jackson mutinied, which compelled its surrender and led to the surrender of Fort St. Philip.

Having captured New Orleans, Flag Officer Farragut took his flotilla farther upriver where it seized Baton Rouge and Natchez. The flotilla remained inland cooperating with the Federal Army until after the surrender of Vicksburg in July 1863.

TABLE 8-3: JANUARY-APRIL 1862	Attempts	Successes	Losses
North Carolina (Wilmington)	5	5	0
South Carolina (Charleston)	18	17	1
Georgia (Savannah) and East Florida	6	6	0
West Florida (St. Marks)	1	1	0
Alabama (Mobile)	0	0	0
Louisiana (New Orleans)	21	16	5
Texas (Galveston)	2	2	0
Lost at Sea	2	-	2
Totals	55	47	8

Timeline: The ironclad CSS *Virginia* began its attack on Federal vessels in the Chesapeake Bay on March 8, 1862, and the battle with the USS *Monitor* took place on March 9. The Peninsular Campaign began in March 1862. The Battle of Shiloh occurred on April 6-7, 1862.

During the period January through April 1862 (see Table 8-3) the rate of blockade running rose to one attempted voyage every other day, and it remained at that level for the rest of the year. Initially, the principal destinations were Louisiana (21 attempts, 16 successes) and Charleston (18 attempts, 17 successes), and the secondary destinations were Georgia (six attempts, all successful) and Wilmington (five attempts, all successful). The Federal blockaders had occupied the Mississippi River below Forts Jackson and St. Philip, and all the steam vessels arriving at Louisiana used secondary destinations—Grand Caillou, Brashear City and Barataria Bay—rather than New Orleans. Of the seven attempts that blockade runners made to depart the Confederacy from points on the Mississippi River, only three were successful, and the other four vessels were either captured or turned back. Most of the successful departures were from Brashear City. New Orleans was the northern terminus of a railroad to Brashear City, and, in early May 1862, the Federals entered Brashear City, which ended blockade running there. The Federals' occupation of Tybee Island and operation against Fort Pulaski closed the main access to Savannah, so all arrivals and departures of steam-blockade runners in Georgia and east Florida occurred at New Smyrna.

During the period May through August 1862 (see Table 8-4), over half the attempted voyages were made at Charleston (30 attempts, 21 successes). The nine failed attempts at Charleston coincided with Commander Marchand becoming senior officer of the blockaders off Charleston in May and the improved effectiveness of blockading there. No vessels arrived at or departed from Louisiana, and all other destinations—including west Florida and Texas—had attempts and successes in the single digits.

THE CAMPAIGN: 1861-1863

TABLE 8-4: MAY-AUGUST 1862	Attempts	Successes	Losses
North Carolina (Wilmington)	4	3	1
South Carolina (Charleston)	30	21	9
Georgia (Savannah) and East Florida	6	3	3
West Florida (St. Marks)	5	4	1
Alabama (Mobile)	6	3	3
Louisiana (New Orleans)	0	0	0
Texas (Galveston)	4	3	1
Lost, Destroyed or Captured at Sea	1	-	1
Totals	56	37	19

Timeline: On May 31, 1862, General Joseph E. Johnston was wounded at the Battle of Seven Pines. General Robert E. Lee took command of the Confederate forces before Richmond the next day. In the Seven Days Battle, June 25 to July 1, 1862, General Lee forced the Federal Army away from Richmond. The Second Battle of Bull Run occurred on August 29-30, 1862.

TABLE 8-5: SEPTEMBER-DECEMBER 1862	Attempts	Successes	Losses
North Carolina (Wilmington)	4	4	0
South Carolina (Charleston)	22	18	4
Georgia (Savannah) and East Florida	0	0	0
West Florida (St. Marks)	1	1	0
Alabama (Mobile)	3	3	0
Louisiana (New Orleans)	1	0	1
Texas (Galveston)	2	1	1
Lost at Sea	1	-	1
Totals	34	27	7

Timeline: On September 4, 1862, General Lee began the Confederate invasion of Maryland. The Battle of Antietam occurred on September 17, 1862. President Lincoln issued the preliminary Emancipation Proclamation on September 22, 1862. The Battle of Fredericksburg occurred December 13, 1862.

The period September through December 1862 (see Table 8-5) saw a slight reduction of the attempts to run the blockade in general and at Charleston (22 attempts, 18 successes), probably as a result of capturing or destroying blockade runners there during the prior period. The fewer captures at Charleston probably reflected the increased care and stealth exercised by the remaining blockade runners. Despite the reduction, more than 60 percent of all attempts at blockade running remained concentrated at Charleston. Otherwise, the period saw little change in the level of attempts or the centers of blockade-running activity.

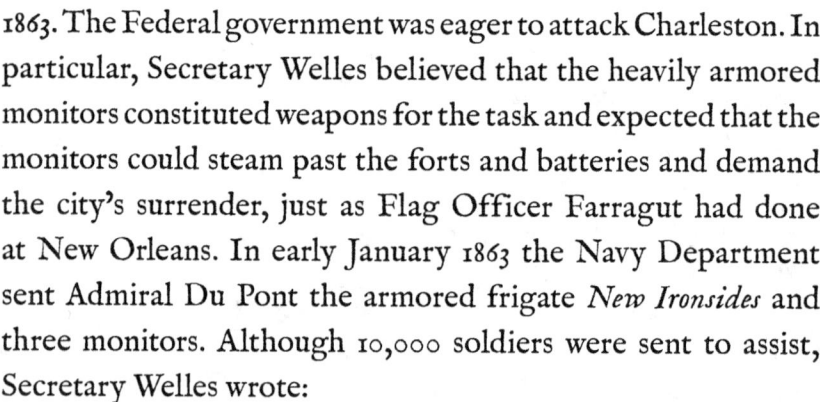

1863. The Federal government was eager to attack Charleston. In particular, Secretary Welles believed that the heavily armored monitors constituted weapons for the task and expected that the monitors could steam past the forts and batteries and demand the city's surrender, just as Flag Officer Farragut had done at New Orleans. In early January 1863 the Navy Department sent Admiral Du Pont the armored frigate *New Ironsides* and three monitors. Although 10,000 soldiers were sent to assist, Secretary Welles wrote:

> The capture of this most important port, however, rests solely upon the success of the naval force, and it is committed to your hands to execute, with the confidence which the Department reposes in your eminent ability and energy.

Secretary Welles stated his desire for a naval attack upon Charleston but placed on Admiral Du Pont the decision whether it could be done. Secretary Welles forwarded a War Department memorandum that critiqued the defenses at Charleston and recounted a conversation with the Confederate commandant of Fort Sumter who said that unless a vessel could be detained for 10 minutes in the channel between Fort Sumter and Fort Moultrie, "nothing could stop the capture of Charleston by vessels determined to run his fire." And although the Confederates attempted to block the channel, the current swept the obstructions away.

By March the Navy Department had increased Admiral Du Pont's ironclad flotilla to nine vessels, all the time encouraging him to attack Charleston and then Savannah. In early April, Secretary Welles wrote to Admiral Du Pont, saying:

> Matters are at a standstill on the Mississippi River, and the President was with difficulty restrained from sending off Hunter and all the ironclads directly to New Orleans, the opening of the Mississippi being considered the principal object to be attained. It is, however, arranged as you will see by to-day's order, that you are to send all the ironclads that survive the attack upon Charleston immediately to New Orleans, reserving to your squadron only two. We must abandon all other operations on the coast where the ironclads are necessary to a future time. We can not clear the Mississippi River without the ironclads, and as all supplies come down to Red River, that stretch of the

river must be in our possession. This plan has been agreed upon after mature consideration and seems to be imperative.

Admiral Du Pont had his doubts. Having used gunboats and monitors singly and in groups to attack Fort McAllister in Ossabaw Sound, he had developed a strong impression of the ironclads' abilities, limitations and vulnerabilities. He acknowledged their defensive capabilities, but he disparaged their limited offensive powers. The Federal government's enthusiasm for both attacking Charleston and the monitors deafened it to his doubts and reservations. Indeed, his awareness of that enthusiasm may have dissuaded him from expressing himself candidly since to do so could have cost him his command.

Forts and shore batteries defended Charleston Harbor from a naval attack and formed three circles of fire through which the attackers would have to pass to reach the city. The Confederates instructed their gunners to aim for the waterline of wooden vessels and to target the decks of ironclads or the junction of the turret with the deck—the Confederates had information that deck armor was thin, and they probably supposed that hitting the base of the turret would disable it. They also placed obstructions and torpedoes to hamper the passage of Federal vessels into the inner harbor. Long guns were to concentrate on the leading vessels to disrupt the procession so as to keep all the vessels under fire for a longer time. Mortars were ordered to aim their fire at the center of the circle of fire and to use full fuses so that the shells would not burst in the air. If the attacking fleet was large, the mortars would continue to target the center of the first circle. If it was not large and a portion had passed to the inner harbor, certain mortars were to redirect their fire at the inner harbor. Within the inner harbor, the Confederates stationed a flotilla of 15 rowed boats—skiffs, canoes and cutters—armed with 60-pound torpedoes mounted on 20-foot spars. Torpedo boats had not been used in the war thus far, and the Confederates hoped that the intrinsic weakness of rowed boats would be overcome by the element of surprise.

The Federals expected to encounter obstructions and torpedoes in the harbor, and the fact that they were covered by Confederate guns made them difficult to counter. John Ericsson, the designer of the *Monitor*, proposed to deal with both floating obstacles and underwater obstructions by building a raft to be attached to the bow of the lead monitor. Grappling hooks hanging below the raft were intended to snag any ropes and floating torpedoes and hold them away from the vessel. A shell holding several hundred pounds of gunpowder, to be detonated with a percussion primer, was suspended 12 feet below the raft to blast any obstruction it encountered. The Federal officers feared that the shell, even if it worked as intended, could damage the monitor that carried the raft at its bow. Moreover, the currents within the harbor were tricky; the monitors were sluggish at best and attaching the raft to a monitor made it more difficult to handle; and under combat conditions, the presence of the shell made the possibility of a collision with another Federal vessel all the more dangerous. The commander of the *Weehawken* consented to use the raft, but he refused to use the shell unless upon a direct order.

The nine Federal ironclads weighed anchor around noon on April 7, 1863. The order of attack was four monitors in the lead, followed by the *New Ironsides* and three more monitors and the *Keokuk*, an experimental vessel with two fixed turrets and armor composed of layers of wood and iron bars. The grappling hooks on the raft attached to the lead monitor became tangled with the anchor chain, which delayed the advance until a quarter past one. The ironclads passed a number of buoys that raised a suspicion that torpedoes were present. Confederate accounts differ as to whether any torpedoes had been deployed. Something exploded as the lead monitor passed, but it did not do any damage. About 10 minutes after two the lead monitor signaled the presence of obstructions, which appeared to be several rows of casks strung between Fort Moultrie and Fort Sumter. Rows of pilings were seen beyond the casks. About 10 minutes before three, the guns of Fort Moultrie opened, and the other guns around the outer harbor joined in. The ironclads stopped their forward motion and became entangled with one another. The *New Ironsides* let down her anchor twice to avoid running aground. The monitors and the *Keokuk* fired into Fort Sumter at ranges from 550 to 800 yards, and the *New Ironsides* fired from a distance of about 1,000 yards. At half past four Admiral Du Pont signaled to withdraw from action.

The intense firing lasted about 40 minutes. During the attack, the Federal vessels, mounting 32 guns, fired 139 shots and shells (16,000 pounds of metal), most at Fort Sumter. The *New Ironsides* was the only Federal vessel capable of rapid firing that might have suppressed the fire from some of the Confederate guns, but she fired only one broadside. The Confederates replied with 69 guns that fired 2,209 shots and shells (162,000 pounds of metal). Confederate engineers counted 55 craters in the northeastern and eastern walls of Fort Sumter. Most were between 1 and 2 feet deep, but a couple were 5 feet deep. The damage was spread across the lower half of the walls, indicating that the Federals failed to concentrate their fire.

Although Admiral Du Pont said he intended to renew the attack the next day, he changed his mind when he learned of the condition of his ironclads. The *Keokuk*'s armor provided little protection—her hull was breached at or below the waterline in 19 places, and she sank the next morning. (The Confederates recovered her guns.) The other vessels showed varying amounts of damage. Hits on the turret or pilothouse sometimes sent nuts or bolt ends flying within. The turrets were dented but not breached. At least one monitor suffered a hole in her deck plating. Hard use of the monitors' equipment caused breakage that interrupted their ability to fire their guns: several turrets were jammed, a port stopper was jammed shut and a gun carriage broke during firing. Some of the injuries were repaired quickly, while others took more time. Although hit many times, the Federal armored vessels generally remained able to fire their guns. The artist Xanthus Smith served aboard a Federal warship near Charleston. His drawing of the monitors after the attack showed some of the external damage they suffered, most of which was to the smokestacks, but these were not armored, and, because blowers provided ventilation, such damage was of little consequence. Human casualties were light: the Federals and the Confederates each suffered one killed and several wounded.

Admiral Du Pont became more vocal in his criticism of the monitors. An ardent proponent of the monitors observed that they sustained little damage and, with a little repair work, would be able to renew the attack, improved by what had been learned in combat. Secretary Welles, having invested much of his budget and his personal political capital in the monitors, said he regretted that Admiral Du Pont had not been more direct prior to the attack in expressing his doubts about the monitors' capabilities and the wisdom of attacking the fortifications. Admiral Du Pont was replaced.

TABLE 8-6: JANUARY-APRIL 1863	Attempts	Successes	Losses
North Carolina (Wilmington)	38	35	3
South Carolina (Charleston)	50	41	9
Georgia (Savannah) and East Florida	4	4	0
West Florida (St. Marks)	3	3	0
Alabama (Mobile)	12	12	0
Louisiana (New Orleans)	0	0	0
Texas (Galveston)	3	3	0
Lost, Destroyed or Captured at Sea	4	-	4
Totals	114	98	16
Timeline: President Lincoln issued the Final Emancipation Proclamation on January 1, 1863. On April 29, 1863, Federal forces under General Ulysses S. Grant conducted a landing on the banks of the Mississippi River below Vicksburg that led to the capture of that city.			

The start of 1863 (see Table 8-6) saw an enormous increase in blockade running as the Confederates abandoned King Cotton diplomacy—withholding their cotton in an attempt to force diplomatic recognition—and began to sell cotton to pay for imports. The rate of blockade running rose to one attempted voyage per day. The principal destination remained Charleston (50 attempts, 41 successes) with nearly 44 percent of all attempts made in the Confederacy, but Wilmington (38 attempts, 35 successes) increased in importance with about 33 percent. Another 10 percent were made at Mobile (12 attempts, all successful).

The increased number of blockade runners taxed the blockaders, even at Charleston where they had been having their greatest success. When Commander Marchand became senior officer at Charleston, the blockade running failure rate at Charleston rose to 30 percent during May through August 1862; it fell to around 18 percent during September through December 1862; and it remained there during January through April 1863. Despite the decline, the blockading at Charleston remained more effective than at any other port that remained open—during January through April 1863 the failure rate at Wilmington was under 8 percent and at Mobile it was zero. In September 1862 the Confederate cruiser *Florida* ran into Mobile under false colors. The Federals, embarrassed, tightened the blockade there, but the *Florida* escaped from Mobile in January 1863—again to the embarrassment of the Federals—although neither their embarrassment nor their efforts to tighten the blockade had any practical result.

TABLE 8-7: MAY-AUGUST 1863	Attempts	Successes	Losses
North Carolina (Wilmington)	89	82	7
South Carolina (Charleston)	53	46	7
Georgia (Savannah) and East Florida	2	1	1
West Florida (St. Marks)	5	5	0
Alabama (Mobile)	35	24	11
Louisiana (New Orleans)	0	0	0
Texas (Galveston)	0	0	0
Lost, Destroyed or Captured at Sea	3	-	3
Totals	187	158	29

Timeline: The Battle of Chancellorsville occurred on May 1-4, 1863. The Battle of Gettysburg occurred on July 1-3, 1863. After an extended siege, Confederate forces in Vicksburg surrendered on July 4, 1863. The draft riots in New York City occurred on July 13-16, 1863.

During the period May through August 1863 (see Table 8-7) the rate of blockade running increased to two attempted voyages every three days. A substantial increase in blockade running at Wilmington made it the leading destination on the Confederate coast (89 attempts, 82 successes) with nearly half the attempts made in the Confederacy as a whole. The unsuccessful ironclad attack upon Fort Sumter at Charleston in April 1863 had little effect upon the level of blockade running there (53 attempts, 46 successes). The attempts to run the blockade at Mobile nearly tripled from the prior period (35 attempts, 24 successes), but the failure rate increased to just over 30 percent. Commander Marchand was reassigned to the West Gulf Blockading Squadron and joined the blockading force off Mobile in March 1863, and this dramatic increase in the failure rate was probably due to his direction of the blockading forces there.

The appointment of Rear Admiral John A. Dahlgren to replace Admiral Du Pont ushered in a period of close cooperation between Federal Army and Navy forces against Charleston.

Morris Island was the long strip of land that defined the southern shore of Charleston's outer harbor. Fort Wagner, near the northern end of the island not far from Fort Sumter, was an irregular earthen fort that stretched across the width of the island, mounting six guns along the land face and three guns along the water face. Battery Gregg, a smaller earthen work that occupied the northern tip of the island, controlled the communications between the island and the mainland. The guns of Fort Wagner and Fort Sumter protected each other.

Federal troops under General Quincy A. Gillmore landed on the southern end of Morris Island in early July 1863, and, after two failed attempts to take Fort Wagner by storm, resorted to siege. Some firing began after the first gun emplacements were built and continued throughout the operations with the support of Federal ironclads and gunboats. The Confederates used the darkness to supply Fort Wagner and repair the damage incurred. The Confederates resupplied Fort Wagner by water and rotated troops in and out at regular intervals. On the night of August 10, the Federals began using calcium lights in earthworks and on vessels to interfere with the Confederate operations. Confederate gunners were unable to extinguish the lights. The lights occasionally interrupted the resupply altogether, but the Confederates switched from larger steamers to smaller boats to keep the supply line open by reducing the chance of detection.

On August 17 the Federals commenced an intense bombardment of Fort Wagner, Battery Gregg and Fort Sumter supported by seven ironclads and seven other gunboats. The damage to Fort Sumter was substantial. Federal troops continued to move their siegeworks closer to Fort Wagner until success in storming it was almost inevitable. Federal troops entered Fort Wagner and Battery Gregg on September 7 and found them deserted. The inevitability of capture persuaded the Confederates to quit Fort Wagner; the calcium lights were prominent in the decision to evacuate; a more thorough use of them might have interrupted the resupply altogether and thereby shortened the siege.

The near wall of Fort Sumter had been reduced to rubble. The Federal Navy's attempt to capture Fort Sumter by landing men in small boats on the night of September 8 to 9, 1863, was a disaster. Out of 400 attackers, about 117 men were killed or captured. Admiral Dahlgren remained eager to launch a monitor attack against the inner harbor of Charleston. He estimated that 10 monitors would be needed, and he experimented with the raft and the underwater gunpowder charge to remove obstructions. The Federal Navy Department cooled to the idea of continued operations at Charleston after the failed boat attack. At the start of 1864 a portion of the Federal troops in South Carolina were reassigned to Virginia, and, although Federal operations against Charleston and its defenses did not cease, they had little effect.

TABLE 8-8: SEPTEMBER-DECEMBER 1863	Attempts	Successes	Losses
North Carolina (Wilmington)	134	119	15
South Carolina (Charleston)	2	1	1
Georgia (Savannah) and East Florida	6	1	5
West Florida (St. Marks)	23	20	3
Alabama (Mobile)	4	1	3
Louisiana (New Orleans)	0	0	0
Texas (Galveston)	0	0	0
Lost, Destroyed or Captured at Sea	2	-	2
Totals	171	142	29
Timeline: The Battle of Chickamauga occurred on September 19-20, 1863. President Lincoln delivered the Gettysburg Address on November 19, 1863. The Battle of Chattanooga occurred on November 23-25, 1863.			

The presence of Federal vessels within Charleston's outer harbor effectively ended blockade running there, and, during the period September through December 1863 (see Table 8-8) only one steam vessel left Charleston, and none attempted to enter. In August 1863 Admiral Farragut's flotilla, having cooperated in opening the Mississippi River, rejoined the coastal blockade, and during the period blockade running at Mobile fell to almost nothing (four attempts, one success). The vessels engaged in this traffic were mostly riverboats that were not well suited to hauling cargoes across the Gulf of Mexico, and a number of these were captured by September 1863. Wilmington was the major beneficiary of the troubles at Charleston and Mobile. The number of attempts made at Wilmington, already large, increased to about 78 percent of all attempts at blockade running made in the Confederacy (134 attempts, 119 successes). St. Marks, Florida, was a secondary beneficiary with over 13 percent of all attempts (23 attempts, 20 successes). All other ports accounted for about 9 percent of all attempts.

9. The Campaign: 1864-1865

TABLE 9-1: JANUARY-APRIL 1864	Attempts	Successes	Losses
North Carolina (Wilmington)	115	101	14
South Carolina (Charleston)	7	4	3
Georgia (Savannah) and East Florida	0	0	0
West Florida (St. Marks)	4	2	2
Alabama (Mobile)	18	18	0
Louisiana (New Orleans)	0	0	0
Texas (Galveston)	11	9	2
Lost, Destroyed or Captured at Sea	3	-	3
Totals	158	134	24
Timeline: President Lincoln appointed General Grant to the command of all the Federal Armies on March 9, 1864.			
Note: The statistical information in Tables 9-1 to 9-4 is derived from S.R. Wise, *Lifeline of the Confederacy* at 233-284.			

The period January through April 1864 (see Table 9-1) brought some changes in the pattern of blockade running. The traffic remained large at Wilmington (115 attempts, 101 successes) with nearly 73 percent of the attempts. Preparations for military operations in Virginia and Georgia pulled forces away from Charleston, and the reduction of military operations there led to a modest resumption of blockade running (seven attempts, four successes). Blockade running at Mobile resumed double-digit levels (18 attempts, all successful). Purpose-built vessels employed to run the blockade at Mobile proved more effective than the river steamers. Blockade running to Galveston and nearby destinations increased (11 attempts, 9 successes)—only 12 attempts had been made to reach or leave Texas destinations in all the prior periods. The Federals occupied Galveston in October 1862, and, while the Confederates recaptured the town in January 1863, blockade running there did not pick up until January 1864. Attempts involving St. Marks and other west Florida destinations dropped from 23 from September through December 1863 to 4 from January through April 1864 (2 successes). Only four vessels were responsible for the 23 voyages during the prior period; three were destroyed or captured in January and February 1864; and the fourth was sent to run the blockade at Wilmington.

1864. Circumstance gave Mobile a two-year reprieve. When Secretary Welles sent Admiral Farragut to the Gulf of Mexico in 1862, his orders were to capture New Orleans and then take Mobile, but Federal operations kept Admiral Farragut and his flotilla on the Mississippi River. Major General Nathaniel P. Banks, commander of the Federal Department of the Gulf, had instructions to move against Mobile once Federal forces had secured the Mississippi River, but after Vicksburg and

Port Hudson fell in July 1863, General Banks was ordered to Texas. Upon taking command of all the Federal Armies in March 1864, General Ulysses S. Grant ordered simultaneous and continuous attacks on Confederate Armies and resources. Although General Banks and Admiral David D. Porter already had embarked on an expedition up the Red River into western Louisiana, General Grant ordered General Banks to attack Mobile even if it meant "the abandonment of the main object of the Red River expedition." Military reversals impeded the expedition, low water stranded the Federal flotilla and only the presence of the Federal infantry prevented the Confederates from destroying the Federal vessels. The attack on Mobile was abandoned.

By June 1864 the complexion of the war had changed. A series of battles in Virginia brought a Federal Army south of Petersburg, not far from Richmond, where it pinned a Confederate Army to its trenches. In the west, a Federal Army had started a series of battles that were driving a Confederate Army ever closer to Atlanta. The possibility that Federal attack elsewhere might draw Confederate strength away from Atlanta made Mobile a target once again.

Mobile sat at the northern end of Mobile Bay about 30 miles from the passage to the Gulf of Mexico. The bay was shallow and 15 miles wide at its broadest point. The principal entrance to the bay was about 3 nautical miles wide, defined on the east by Mobile Point and on the west by Dauphin Island, and the main ship channel ran near the eastern edge of the entrance below Fort Morgan, a masonry fort with five sides of roughly equal length and a bastion at each corner. The fort consisted of one tier of casements topped by guns mounted en barbette, but batteries had been constructed in front of the faces of the fort that overlooked the ship channel, and the casements immediately behind them had been sealed.

Fort Gaines sat opposite Fort Morgan on the eastern end of Dauphin Island. It had an earthen rampart with guns mounted on top en barbette surrounded by a Carnot wall—a scarp wall that stood free of the rampart and rose just high enough for the guns to be fired over it. The Carnot wall prevented the fort from being taken by a coup de main. When enemy gunfire breached the Carnot wall, the masonry rubble became an obstacle that defended the earthen rampart within.

The Confederates obstructed the western portion of the main entrance to the bay with a line of pilings. Closer to the ship channel, the Confederates had placed a line of obstructions consisting of buoys with ropes to interfere with paddlewheels and snag propellers. Sixty-seven torpedoes were mixed among the floating obstructions or set adjacent to them, although most had been in the water for a long time and may have broken free or been neutralized by seawater. The Confederates had a reserve of about 20 torpedoes to be placed when Federal gunboats threatened to attack. They also had nine torpedoes that were detonated by electricity, but only three of these had been placed.

Several shallow passes west of Dauphin Island also offered entry into Mobile Bay. Fort Powell, a sand fort, guarded Grant's Pass, but the work remained incomplete by August 1864. Earlier in the year, Admiral Farragut had attacked Fort Powell with six mortar boats, but two weeks of bombardment made little impression upon it. The Confederates perceived the attack as a mere demonstration, noting that each day's bombardment ceased at dark, and they repaired the damage to the fort and resupplied it overnight.

The four Confederate warships in Mobile Bay included the *Tennessee*, an ironclad that mounted six rifled cannon, and three side-wheel steamers that each mounted up to six cannon. The Federals did not have ironclads on hand to oppose the *Tennessee*, and Admiral Farragut instructed his commanders to use desperate measures if she came out into the Gulf: "No commanding officer will err in risking his vessel by running down the enemy, but the destruction of the ironclad *Tennessee* should be the aim of all."

The Federals took steps to counter the threat posed by the *Tennessee*. They mounted curved iron beaks called "cutters" on the bows of their larger warships. The cutter extended 2 feet above the waterline and 3 feet below, with the concave surface facing forward. The cutter might pierce the hull of a rammed ship, and saw teeth on the forward edge were intended to cut rope obstructions. The Federals also attached iron plates to the bow of at least one warship. Admiral Farragut obtained torpedoes, and, although he regarded the weapons as "unworthy of a chivalrous nation," he concluded that "it does not do to give your enemy such a decided superiority over you." Having read in a newspaper that steel shot was more effective against iron armor, he requested some from the Navy Department, but it is not clear that he received any.

By the eve of battle, four monitors joined the Federal squadron. The USS *Manhattan* and the USS *Tecumseh* were *Canonicus*-class monitors. The USS *Chickasaw* and the USS *Winnebago* were double-turreted river monitors that had been built for service on inland waters.

The Federal plan of attack had the four monitors lead 14 wooden gunboats, lashed together in pairs, up the main ship channel. Upon reaching Fort Morgan, the monitors would stop and engage the fort. While approaching the fort, only the forward pivot guns on the wooden gunboats would be able to fire on the fort. As each gunboat drew abreast of the fort behind the monitor screen, she would turn and fire her broadside guns. Once they had turned into the bay, the gunboats would be able to fire only their aft pivot guns. The pairs consisted of a larger propeller-driven boat paired with a smaller side-wheeler. The larger vessels were placed closest to the fort; they mounted more guns and, being propeller-driven, were less vulnerable to being disabled by enemy fire. Admiral Farragut chose the USS *Brooklyn* to be the lead vessel because she possessed a torpedo catcher, a device with rods or hooks, shaped like a locomotive's cowcatcher, that was intended to snag or detonate torpedoes away from the hull.

The attack began on August 3, 1864, when Federal troops landed on Dauphin Island and invested Fort Gaines. By midnight on August 4, they had moved their light artillery to a position 1,200 yards from Fort Gaines, and, on the morning of August 5, as the Federal warships started to move, the Federal artillery opened fire on the fort, attempting to silence its guns facing the water as the warships passed. Fort Morgan opened fire as the warships approached. The lead monitor *Tecumseh* failed to remain in the ship channel, passed into the obstructions and struck a torpedo that exploded. The *Tecumseh* sank rapidly. The rest of the Federal warships, following the *Tecumseh*'s lead and poised to proceed through the obstructions and torpedoes, slowed their progress. From the *Hartford*, immediately behind the *Brooklyn,* Admiral Farragut ordered the *Brooklyn* to "go ahead," and then the *Hartford* steered around the *Brooklyn* and led the squadron into the bay. The men aboard the Federal gunboats heard the triggering devices of the Confederate torpedoes click and snap as they passed through the obstructions, but none exploded. Inside the bay, the paired vessels unlashed. The *Tennessee* attempted to ram the *Hartford* but failed. Next, the *Tennessee* attempted to ram the *Brooklyn,* but the *Brooklyn* either avoided her or the *Tennessee* turned away—the *Brooklyn* had lifted her torpedo catcher, which may have appeared to be a spar torpedo. All the Federal vessels had passed into the bay by half past eight. The *Tennessee* retreated under the guns of Fort Morgan, and the Federal vessels anchored. Shortly after, the Federals saw the *Tennessee* bearing down upon them. Several Federal vessels rammed the *Tennessee,* and a number of them mobbed around her and fired into her at point-blank range. The monitor *Chickasaw,* astern of the *Tennessee,* battered her severely with an 11-inch smoothbore gun. A shot from a 15-inch smoothbore gun on the monitor *Manhattan* broke the *Tennessee*'s armor and sent splinters into the casement, although the shot did not fully pierce the *Tennessee*'s side. During the last 30 minutes of the battle, the *Tennessee*'s guns did not reply to the Federal fire. Finally, the *Tennessee* surrendered. The Federals captured one Confederate gunboat soon after they entered the bay, and two others succeeded in getting under the guns of Fort Morgan. One sank and the other escaped up the bay to Mobile.

The Federals did not succeed in silencing the guns of Fort Morgan, which fired 491 rounds. The *Hartford* and the *Brooklyn* suffered most of the damage from the fort's guns. The Federal squadron suffered additional damage from the Confederate flotilla, especially the *Tennessee*, whose guns were fired at close range. The *Tecumseh* had 114 men on board, but only 21 escaped after she struck the torpedo. The other casualties in the Federal gunboats included 52 killed and 170 wounded. Aside from the *Tecumseh*, the monitors did not experience any casualties.

On the same morning, five of the Federal gunboats that remained in the Gulf of Mexico shelled Fort Powell at long distance. Fort Powell mounted eight guns. The front face of Fort Powell was complete and mounted four cannon, and the side face nearest to Fort Gaines, although only half complete, mounted two guns. Shots from the Federal gunboats in the Gulf of Mexico hit the fort five times but without doing any appreciable damage. At 2:30 that afternoon one of the monitors in the bay approached to within 700 yards of Fort Powell and began firing shell and grape shot. Fort Powell replied with shots from a rifled cannon. Once darkness had fallen, the Confederates spiked their guns and blew up the fort.

The siege of Fort Gaines continued. On August 6, the double-turreted monitor the *Winnebago* fired on the fort. On August 8, the Federal battery of four rifled cannon and six field guns opened fire, and Fort Gaines surrendered two days later. While the fort had abundant ammunition and food, it was deficient in the protection it provided. Federal troops, guns and gunboats then moved to operate against Fort Morgan. They invested Fort Morgan and commenced siege operations. Fort Morgan surrendered on August 23.

Admiral Farragut took a gamble by entering Mobile Bay before Fort Morgan had been captured, similar to the risk he ran in proceeding past the forts up the Mississippi River. The

Confederates' abandonment of Fort Powell opened communication through Grant's Pass into the bay. The Confederate commander at Mobile complained about the abandonment of Fort Powell, noting that the soldiers manning the guns "would have been exposed as are sailors on an ordinary man-of-war." The comparison was not apt: the defenders of Fort Powell might have persevered to drive off a small wooden gunboat, but they would not have driven off a monitor firing on them methodically from close at hand.

TABLE 9-2: MAY-AUGUST 1864	Attempts	Successes	Losses
North Carolina (Wilmington)	141	129	12
South Carolina (Charleston)	22	19	3
Georgia (Savannah) and East Florida	1	0	1
West Florida (St. Marks)	0	0	0
Alabama (Mobile)	21	19	2
Louisiana (New Orleans)	0	0	0
Texas (Galveston)	12	11	1
Lost, Destroyed or Captured at Sea	1	-	1
Totals	198	178	20
Timeline: In May 1864 General Grant began the Overland Campaign toward Richmond and General William T. Sherman began his advance upon Atlanta. On September 2, 1864, General Sherman captured Atlanta.			

The period May through August 1864 (see Table 9-2) saw relatively small changes in the pattern of blockade running. Wilmington remained the principal destination (141 attempts, 129 successes) with 71 percent of the attempts, and the activity at Charleston picked up a little (22 attempts, 19 successes). Blockade running at Mobile remained modest (21 attempts, 19 successes) until the Federals captured Mobile Bay, which ended blockade running there. The amount of blockade running through Galveston stayed about where it had been during the prior period (12 attempts, 11 successes).

TABLE 9-3: SEPTEMBER-DECEMBER 1864	Attempts	Successes	Losses
North Carolina (Wilmington)	117	100	17
South Carolina (Charleston)	45	41	4
Georgia (Savannah) and East Florida	0	0	0
West Florida (St. Marks)	0	0	0
Alabama (Mobile)	0	0	0
Louisiana (New Orleans)	0	0	0
Texas (Galveston)	18	16	2
Lost, Destroyed or Captured at Sea	0	-	0
Totals	180	157	23
Timeline: On November 8, 1864, Abraham Lincoln was elected to a second term as president. On November 15, 1864, the Federal Army under General Sherman began its march to the sea. The Battle of Nashville occurred on November 15, 1864. General Sherman's Army reached Savannah on December 21, 1864.			

Only Wilmington, Charleston and Galveston remained open during the period from September through December 1864 (see Table 9-3). Blockade running decreased slightly at Wilmington (117 attempts, 100 successes) with 65 percent of the attempts. Blockade running doubled at Charleston (45 attempts, 41 successes), which rose to 25 percent of the attempts, and increased at Galveston (18 attempts, 16 successes) to 10 percent of the attempts.

Before the Civil War began, Wilmington was the most active port in the state of North Carolina, known mostly for shipping naval stores—tar, turpentine and related products made from the sap of pine trees—although it was minor in comparison to the principal cotton ports of the South. The port sat about 25 miles up the Cape Fear River that had two separate inlets from the ocean. The southernmost, called the Old Inlet, was guarded by Fort Caswell, a masonry fort. The Confederates built substantial earthworks that protected its masonry faces and additional earthworks at the mouth of the Old Inlet. The New Inlet was located 5 miles east and north of the Old Inlet, but the two inlets were separated by Frying Pan Shoal that extended 20 miles out into the ocean. The Federal blockading vessels patrolled the two inlets as if they served different ports.

The ship channel leading to the New Inlet ran parallel to the shore from the north of the inlet past a long, sandy peninsula known before the war as Federal Point. When the Confederates took possession, they renamed it Confederate Point. By 1864 the Confederates had built a massive L-shaped earthwork called Fort Fisher. The short base of the L faced the land to the northeast above Fort Fisher, stretching 480 yards across the peninsula from near the shore to a swamp that bordered the Cape Fear River. The long vertical stroke of the L paralleled the shore for a distance of approximately 1,300 yards and guarded the ship channel. The sand walls of Fort Fisher rose about 30 feet in the air. The end of the water face nearest New Inlet terminated in a sand mound that rose over 40 feet in the air and mounted two large guns, called the Mound Battery. The Confederates installed lights there as a navigational aid and a signal to blockade runners. Battery Buchanan, mounting four large guns, sat below Fort Fisher at the extreme southwestern end of Confederate Point. Ships entering the Cape Fear River through the New Inlet moved under the 23 heavy guns on the water face of Fort Fisher as well as the four guns at Battery Buchanan. The armament on the land face included 20 heavy guns and several field guns and mortars. A 9-foot-tall palisade of logs stood in front of the land face of the fort. The area beyond the palisade had been planted with 24 torpedoes (landmines) that could be detonated by an electrical charge.

For all of its size and apparent strength, Fort Fisher was only half a fort—it lacked rear walls, and its only nod to the defense of the rear was an isolated battery and two lines of rifle pits. The Federals apparently were not aware of this vulnerability: their December 1864 report on Fort Fisher referred to it as an irregular four-sided fort.

In late September 1864 General Grant told General Godfrey Weitzel that he was to command a joint Army-Navy expedition to close the port of Wilmington. The orders were passed through General Benjamin F. Butler, who commanded the district of Virginia and North Carolina. The preparations became so widely known that General Grant postponed the operation. General William T. Sherman, having captured Atlanta, abandoned his supply line and marched his Army eastward toward the sea in November. Shortly after, Confederate President Jefferson Davis ordered General Braxton Bragg—then commanding at Wilmington—to proceed in force to Augusta, Georgia, to oppose General Sherman's Army. When General Grant read newspaper reports of General Bragg's movements, he advised General Butler of the renewed opportunity to move against Fort Fisher and Wilmington, and he urged speedy action. He sent a similar message to Rear Admiral David D. Porter who was to command the Federal Navy forces in the attack. Admiral Porter replied that he could start in three days, but arrangements for "the powder ship" would take "a little longer."

The idea of using a ship loaded with gunpowder to damage or destroy the fort, once proposed, took on a life of its own and controlled the progress of the Fort Fisher expedition. By mid-November Federal engineering and ordnance officers were discussing whether the attempt would work and how the explosion should be detonated. General Grant was growing concerned about the delay, and, on December 4, 1864, he wrote to General Butler:

> I feel great anxiety to see the Fort Fisher expedition off, both on account of the present fine weather, which we can expect no great continuance of, and because Sherman may now be expected to strike the sea-coast any day, leaving Bragg free to return. I think it advisable for you to notify Admiral Porter and get off without delay, with or without your powder boat.

He wrote to General Butler again on December 6, but neither the Army nor the Navy shared General Grant's sense of urgency. General Grant sent further messages to General Butler on December 7, December 11 and December 14. By early December the Navy had acquired a ship and was loading her with gunpowder, and by mid-December the preparations were complete. The Federal troops did not embark until December 14, and the Federal gunboats were not prepared until a few days later. By then, the weather had worsened, and a storm prevented the expedition from sailing.

The Federal gunboats began assembling near Wilmington on December 20, 1864. Even though the Federal troops had not arrived, Admiral Porter commenced operations on December 23 with the powder ship. The powder ship had an iron hull with a flat bottom. The original plan was to run the powder boat onto the beach near the fort, but the possibility that a leak would foul the gunpowder caused the ship to be anchored several hundred yards from the corner of Fort Fisher where the land and sea faces met. Timers were set, and the ship was set on fire as a backup means to detonate the gunpowder. The explosion occurred about 1:40 in the morning of December 24. Observers on the gunboats saw a bright flash followed by four distinct reports. The explosion did not damage the fort, and the Confederates did not perceive it as an attack. A Confederate report said, "A gun-boat last night in pursuing one of our steamers grounded, was abandoned and blown up."

The barrage of Fort Fisher began at half past 11 the same morning. The *New Ironsides* and four monitors led a squadron of 64 vessels into position—the ironclads about three-quarters of a mile from the fort, and the larger ships about a mile from the fort. The warships fired at the rate of 115 rounds per minute. During the day's bombardment, five 100-pounder Parrott rifles on the Federal gunboats burst, killing or wounding 45 sailors. The firing continued until about 5:30 that evening when the Federal gunboats retired for the night. Although Admiral Porter claimed to have silenced the guns of the fort, the Confederates had fired 672 rounds and hit two of the Federal vessels. Confederate losses for the day were 23 wounded.

Federal fire damaged several guns in the fort, and two Brooke rifles in the fort had burst. The Confederates repaired most of the damage to the fort overnight.

The transports with the Federal infantry arrived during the afternoon of December 24 with General Butler in command. The gunboats renewed their bombardment at seven the next morning from positions slightly closer to the fort, and firing continued for about seven hours, concentrating on the water face. The guns of the fort again replied slowly, firing 718 rounds at the gunboats. Admiral Porter considered sending light steamers into the Cape Fear River, and he sent several boats to mark the channel through the New Inlet, but the Confederates fired upon them and drove them away.

The Federal infantry began landing about noon, and roughly 2,300 men got ashore about 5 miles from the fort. They advanced to about 75 yards from the land face of the fort and established a skirmish line. The 17 guns on that face remained silent during the bombardment, but they also were undamaged. Most of the garrison remained in the bombproofs while the shelling continued, but, when it ceased, the Confederate garrison emerged and fired on the Federal troops with muskets and cannon loaded with grape, canister and shell. By five that evening, General Butler had decided to withdraw the troops. General Butler reported that after a "thorough reconnaissance" he and General Weitzel were of the opinion that Fort Fisher "could not be carried by assault as it was substantially uninjured as a defensive work by the Navy fire." Believing that nothing but a regular siege would subdue the fort and noting that a siege was not a part of his instructions, General Butler withdrew, stating that he could "see nothing further that can be done by the land forces." Federal casualties on the second day were 19 killed and 64 wounded. The fort had sustained little more damage on the second day's bombardment than on the first. Confederate casualties also were slightly higher: 3 killed and 43 wounded.

General Grant called the aborted attack "a gross and culpable failure." Although General Butler lost his command, he was not solely responsible for the failure. First, the Army and the Navy both lost sight of the need for urgency and permitted the powder boat to become the controlling element of the campaign. Second, the Navy guns had failed to effect substantial damage to the defensive power of the fort. Although they targeted the water face, they did not disable the guns there. When Federal troops landed, they found the guns on the land face undamaged and ready to oppose them. Third, the Army command failed to recognize and seize the opportunities the situation provided. Statements of the landing force suggested an awareness that Fort Fisher was not an enclosed work and was thus vulnerable to an infantry assault.

1865. Notwithstanding the failure, Admiral Porter made known that he was willing to continue operations against Fort Fisher and could work his gunboats nearer to the fort and cover a Federal landing. General Grant selected Major General Alfred H. Terry to command in the second attack on Fort Fisher. The troops were, for the most part, the same as in the first attack. General Grant gave General Terry his initial instructions on January 2, 1865. The Federal troops embarked on January 4 and 5. The transports and Admiral Porter's fleet arrived at the rendezvous off Beaufort, North Carolina, on January 8, and, on January 12, the expedition departed for Wilmington, the lead elements arriving on the same day.

At four in the morning on January 13, Federal vessels were in place to cover the landing on Confederate Point, and, at 8 AM, four Federal ironclads—a double-turreted monitor and three single-turreted monitors—moved to a position just over half a mile from the land face and opened fire. A second line consisting of the *New Ironsides* and 13 additional Federal gunboats moved into a position behind the ironclads, just over three-quarters of a mile from the land face and opened fire. Two additional lines of Federal gunboats, one consisting of 12 vessels and the other consisting of 14 vessels, moved into positions a little under a mile from the water face and opened fire. With the previous experience at Fort Fisher and General Butler's criticism of his squadron still fresh, Admiral Porter gave specific guidance for the exercise of gunnery:

> All firing against earthworks when the shell burst in the air is thrown away. The object is to lodge the shell in the parapets, and tear away the traverses under which the bombproofs are located.... Fire deliberately. Fill the vessels up with every shell they can carry, and fire to dismount the guns and knock away the traverses. The angle [of the fort] near the ships has heavy casements; knock it away. Concentrate fire always on one point.

Orders forbade firing at flags or flagpoles. The Federals were "to pick out the guns."

As the firing began, nearly 200 boats and steam tugs from the Navy arrived at the transports and began the landing, and, by three in the afternoon, nearly 8,000 men with weapons, ammunition, food and entrenching tools were ashore. The Federal troops established a defensive line across Confederate Point to protect their rear. They then moved into positions near the northern end of the land face of Fort Fisher. After a reconnaissance, General Terry decided to take the fort by assault, but siege guns were brought ashore in case the attempt should fail.

At darkness some Federal gunboats retired for the evening, but others maintained their fire throughout the night to confine the garrison to their bombproofs and to prevent them from repairing the damage that had been done during the day. The bombardment was renewed to full vigor the next morning. At the Army's request, the Navy fired on the log palisade.

The Federals scheduled the assault for the afternoon of January 15. Admiral Porter drew from his fleet a "boarding party" consisting of 2,000 men—400 marines armed with carbines and muskets and 1,600 sailors armed with revolvers and cutlasses—to storm the corner of the fort where the two faces met while the Federal Army stormed the northern end of the land face. The marines were to provide covering fire while the sailors climbed the parapet. Although sailors were trained to board enemy vessels and to defend against boarders, none had any experience in land-based operations, nor had the men trained together as a unit.

The Federals had prepared bags of gunpowder to blow up the palisade, but the cannon fire had been effective, and men with axes cleared the way.

At 3:25 Admiral Porter directed the Navy's fire away from the land face. The signal was repeated through the fleet by sounding steam whistles, and the assault began. The Confederates poured out of the bombproofs and into the walls of the fort where they saw the Navy's boarding party and fought them with muskets and the few cannon that remained serviceable. The sailors and marines suffered heavy casualties and were thrown back. In the meantime, Federal troops on the northern end of the land face had climbed the parapet and gained a foothold in the traverses. Other Federal troops moved beyond the land face and confirmed that Fort Fisher was not an enclosed work. They found pits created when the parapets were built and used them for cover. Additional Federal troops entered the fort through the sally port. Hand-to-hand combat continued through the afternoon and into the evening. The torpedoes buried in front of the palisade did not explode. As the light permitted, the *New Ironsides* and the monitors fired into the traverses that were observed to be occupied by the Confederates. At close to 10 that night the Federal troops had won all the traverses, and the remaining Confederates retreated to Battery Buchanan where they surrendered. Federal Army casualties numbered about 955 killed, wounded and missing. Federal Navy casualties, including the boarding party, numbered about 386 killed, wounded and missing. Confederate casualties numbered about 500 killed and wounded and between 1,400 and 1,500 taken prisoner. Federal gunboats entered the Cape Fear River, which closed the port.

In the second attack on Fort Fisher, the Navy created a much better opportunity for the Army, both as a result of the improved gunnery tactics and through the diversion created by the boarding party. The more determined Army leadership took advantage of the opportunity presented. The success of the second attack on Fort Fisher does not indicate that the first attack could have been successful. Too much was different to support such a conclusion. Nonetheless, the second attack benefitted from the first, which served as a training exercise for the second and was the inspiration for the improvements in the second.

TABLE 9-4: JANUARY-APRIL 1865	Attempts	Successes	Losses
North Carolina (Wilmington)	5	2	3
South Carolina (Charleston)	22	19	3
Georgia (Savannah) and East Florida	0	0	0
West Florida (St. Marks)	0	0	0
Alabama (Mobile)	0	0	0
Louisiana (New Orleans)	2	1	1
Texas (Galveston)	45	43	2
Lost, Destroyed or Captured at Sea	0	-	0
Totals	74	65	9

Timeline: The Federal Congress approved the Thirteenth Amendment on January 31, 1865. President Lincoln met with the Confederate peace commissioners in Hampton Roads on February 3, 1865. President Lincoln was inaugurated to his second term of office on March 4, 1865. The Confederates evacuated Richmond on April 2, 1864. President Lincoln toured Richmond on April 4, 1864. General Lee's Army surrendered on April 9, 1864. President Lincoln was shot on April 14, 1865, and he died the next morning.

During the period January through April 1865 (see Table 9-4) blockade running fell to about 40 percent of what it had been in the prior period. The Federals attacked Fort Fisher near Wilmington in December 1864 but failed to capture it. They captured it in January 1865 and closed Wilmington to blockade running. During the period five vessels attempted to reach or leave Wilmington—two vessels successfully departed and three vessels were lost attempting to enter. Galveston (45 attempts, 43 successes) became the principal destination for blockade running with over 60 percent of the attempts. The two additional attempts involved Calcasieu Pass, a place on the Louisiana coast near the Texas border. About 30 percent of the attempts involved Charleston (22 attempts, 19 successes). The Federal Army commanded by General William T. Sherman, having marched from Atlanta to Savannah, approached Charleston as it headed north in mid-February 1865. The Confederates evacuated the city, which ended blockade running there. The capture of Wilmington and Charleston left Galveston as the only blockade-running destination.

Although the Confederate Armies east of the Mississippi surrendered in April and early May 1865, Confederate forces west of the river remained active a while longer. Galveston remained a blockade-running destination into May. The *Lark* and the *Wren*, both British-built steamships, made journeys to and from Galveston during May, and the *Denbigh*, another British-built steamship, ran aground while trying to enter Galveston in May. During May 1865 blockade runners made five attempts to travel to or from Galveston and were successful four times.

On March 9, 1865, the *New York Times* commented on the progress of the campaign for the Confederate coast with the following remarks under the headline "The Seaports of the Confederacy":

> We have practical evidence, from two widely-distant points, of the effect upon the fortunes

of rebeldom of the capture of the last remaining ports of the South. The Richmond papers of Saturday announced the appointment of a special agent by Gen. LEE, to collect whatever in the shape of old muskets, swords, pikes, pistols and sabres could be found in any remnant of the Confederacy to which, he might be able to find access. The steamer Cuba, with news ten days old from Europe, brought word last night that the "Confederate Government" have countermanded large orders for torpedoes. These incidents convey to the world, in a far more pointed form than any official bulletin, the information that the Confederacy is and has been dependent upon importation for arms to make good the large stock that has fallen by capture, at different times, into the hands of our armies. They likewise advertise the complete exclusion of the rebels from commercial intercourse with foreign ports. Torpedoes are no longer a marketable commodity in the Southern market, because they can find no port of entrance and no blockade-runners to take them even as speculative freight. All the British blockade-running ports, from Nova Scotia to New-Providence, are filled with enterprising craft, laden with Confederate ammunition and stores, but unable even to venture within sight of the coast-line of the Confederacy. And the "neutral" interest on both sides of the Atlantic is drooping and despondent to a degree which has no precedent in the history of this war.

Not less pointed than these facts are the latest official returns of the British Board of Trade. For the last-reported month there was a decline in exports of £2,258,963 sterling, as compared with the reports for the same month of the previous year. This doubtless comes in great part of the stoppage of all further Confederate orders for torpedoes, and other non-contraband articles of Southern commerce.

If war had not followed the secession crisis, any reduction of trade between the northern United States of America and the southern Confederate States of America from the severing of economic relations between those nations would have been made up by trade with other nations. But the Civil War followed secession, and with it came the blockade. Under the technological conditions that existed at the time, the Federal blockade was not especially effective in preventing blockade running or capturing determined blockade runners. The primary effect of the blockade, however, was deterrent because the amount of commerce running through Confederate ports not only did not rise upon the commencement of hostilities, but it fell substantially below its prewar levels as seen in the shipments of cotton.

As the market became aware that the Civil War would not end quickly, adventurers with schooners and entrepreneurs with sufficient capital to acquire fast steam vessels accepted the risks to gain the greater profits that cotton commanded—or the increased prices that goods and materiel imported to the Confederacy commanded—both due to the war and the blockade. The amount of cotton exported, and, by extension the amount of goods imported, never reached prewar levels—nor could they inasmuch as cotton cultivation fell—but nonetheless cotton provided the means to obtain and import goods that otherwise would not have been available to support the Confederate war effort.

The analysis of the effects of the blockade is complicated by the fact that cross-border trade between the Federal North and the Confederate South not only was not terminated, but, in some respects, it was encouraged to flourish by government authorities on both sides of the border. And the same economic forces that encouraged blockade running also prompted the growth of cross-border trade. While some military and government officials disparaged the practice as improper, if not immoral, and dispiriting to the men in military service, the opportunity for large profits led others to eager personal participation, and bribery became a common means of completing transactions. Necessity drove Confederate acquiescence in the cross-border trade inasmuch as domestic production and importation could not supply the needs of the military. Federal participation in the trade was less well founded because the results of the trade were directly contrary to the purpose of establishing the blockade. No less than President Lincoln was an active proponent of cross-border trade, and, although he defended it as being "of public importance" and not "merely a concession to private interest and pecuniary greed," some of his arguments for trade were fatuous while other arguments were merely specious. The highly placed and the politically connected sought authorizations to trade through the lines, and in at least one instance President Lincoln was shamed into revoking a privilege he had extended to a political friend. Regulations and policies intended to cause the seizure of cotton from the Confederate government and traitorous individuals and to benefit loyal owners were manipulated to produce the largest possible profit, and Congressional attempts to set policy to restrict the cross-border trade were ignored. Some historians have posited a direct connection between the Federal feeding frenzy over cotton during the Civil War and the corruption of the Gilded Age that followed it—while continuity seems evident, a causal linkage is more speculative. Of course no given age has a monopoly on the mutual encouragement given to one another by wealth, avarice, power and arrogance or their mutual disregard for what the general public might regard as common morality. They always seem to be with us, and only in some historical settings, such as the Gilded Age, did they become an integral part of the standard narrative in the way that a war or a pandemic would dominate another.

The combined effect of the blockade running and the cross-border trade effectively frustrate an attempt to trace the flow of cotton with any accuracy—we can see where it started, and we can see where it ended, but we cannot identify the means or route of transfer. Cotton entered the stream of commerce as a result of blockade running, cross-border trading, captures on

TABLE 9-5: PRODUCTION, CONSUMPTION, EXPORT AND IMPORT (EQUIVALENT 500-POUND BALES)

Year	U.S. Domestic Production	U.S. Domestic Consumption	U.S. Exports of Domestic Cotton*	U.S. Net Imports of Cotton*	British Imports from U.S.	British Imports from West Indies and Nearby
1859	4,309,642	845,410	2,772,937	0	1,756,665	5,726
1860	3,841,416	841,975	3,535,373	0	2,188,434	8,310
1861	4,490,586	369,226	615,032	0	1,528,528	8,051
1862	1,596,653	287,397	10,129	61,731	52,963	15,112
1863	449,059	219,540	22,770	67,695	94,440	16,468
1864	299,372	344,278	23,998	52,405	74,598	41,243
1865	2,093,658	614,540	17,789	68,798	329,804	80,734

Source: Department of Commerce, Bureau of Census, Bulletin 131, *Cotton Production and Distribution*, 1915, 82 (US data). T. Ellison, *The Cotton Trade of Great Britain*, 1886, Table No. 1 (British imports data).

* For the 12 months beginning July 1 of the year specified.

land and at sea and the sale of stocks held from previous years. Due to the extended growing season that the cotton plant requires, we know that almost all American cotton was grown in the Confederacy. Table 9-5 shows the following:

- The 1861 cotton crop suffered little or no effect from the war, but production in subsequent years fell progressively until the war ended in 1865.
- The war reduced United States domestic consumption to less than half of what it had been.
- Cotton exports from the United States fell by over 80 percent in 1861 and then dropped to less than 1 percent of prewar levels for the remaining war years. (The export data reported by the Federal government does not reflect exports from seceded states. Shipments from captured Confederate ports of cotton seized or purchased through the lines may not be included.)
- Although a net exporter of cotton before the war, the United States (Federal portion) became a net importer of cotton. The nominal country of origin varied year to year, but the major sources of this cotton included Britain and Mexico, which probably had been grown in the Confederacy.
- During the war years, British imports of cotton from the United States exceeded the total amount of United States exports, suggesting the difference consisted of cotton run through the blockade and cotton seized or purchased by the Federals and shipped from captured Confederate ports.

- The British West Indies increased its cotton shipment to Britain during the war years. Part of the increase was probably increased production due to the price rise, and at least one authority states that the large increase in 1864 and 1865 was due to blockade running.

Without the blockade, the Confederacy would have had a freer hand to convert its agricultural production into the resources for war. The Confederacy had the industrial capacity to produce only a small fraction of the manufactured goods that it needed to sustain its Armies and civilian population. The Federal proclamation of the intent to impose a blockade ended ordinary commerce between the Confederacy and the nations of the world. Shipping left the Confederate states during May and June of 1861, and the scarcity of goods became noticeable in May. We should note here that the Confederates expected a short war and, for this reason, coupled with the philosophical belief that the hand of government should fall lightly upon the nation, initially did not take steps to prepare early for the long, hard struggle that the war became. With the blockade, the Confederacy did not starve, but it became significantly malnourished, both figuratively and literally. Confederate troops remained on short rations through much of the war, which took a physical toll and affected morale. After the war, General Josiah Gorgas, the Confederate chief of ordnance, observed that while Confederate Armies had never possessed the full supply of munitions that Army regulations called for, supplies had been adequate, and no battle had been lost for want of munitions, and while that might be true, it sounds like a self-justification.

General Edward Alexander Porter, a Confederate artilleryman, in his memoirs praised General Gorgas for supplying the massive amounts of ordnance and agreed that the Confederates did not lose a battle for want of ammunition. But he went on to make a somewhat more candid observation about the state of Confederate arms:

> There is reason to believe that had the Federal infantry been armed from the first with even the breech-loaders available in 1861 the war would have been terminated within a year. The old smooth-bore musket, calibre 69, made up the bulk of the Confederate armament at the beginning, some of the guns, even all through 1862, being old flint-locks. But every effort was made to replace them by rifled muskets captured in battle, brought through the blockade from Europe, or manufactured at a few small arsenals which we gradually fitted up. Not until after the battle of Gettysburg was the whole army in Virginia equipped with the rifled musket. In 1864 we captured some Spencer breech-loaders, but we could never use them for lack of proper cartridges.
>
> Our artillery equipment at the beginning was even more inadequate than our small-arms. Our guns were principally smoothbore 6-Prs. and 12-Pr. howitzers, and their ammunition was afflicted with very unreliable fuses. Our arsenals soon began to manufacture rifled guns, but they

always lacked the copper and brass, and the mechanical skill necessary to turn out first-class ammunition. Gradually we captured Federal guns to supply most of our needs, but we were handicapped by our own ammunition until the close of the war.

Blockading prevented an adequate supply from becoming ample, and blockade running prevented mere adequacy from becoming a dearth. A better-supplied Confederacy would have been stronger and more resilient, and changed conditions might have changed the course of history.

10. Notes

Abbreviations

ORA   The War of Rebellion: A Compilation of the Official Records of the Union and Confederate Armies, Government Printing Office, 1880-1901.

ORN   Official Records of the Union and Confederate Navies in the War of Rebellion, Government Printing Office, 1894-1922.

Chapter 1. Aims and Means

At the end of the first major battle: E.S. Rafuse, A Single Great Victory.

In the aftermath of Bull Run: "Disasters on the Road to Victory", *New York Daily Tribune*, Jul. 25, 1861.

At the time of President Lincoln's: W. Scott to W.H. Seward, Mar. 3, 1861, Lincoln Papers, Library of Congress.

None of the suggestions could: H. Greeley, The American Conflict, Vol. 1 at 359. D.M. Potter, "Horace Greeley and Peaceable Secession", Journal of Southern History, Vol. 7, No. 2 (May 1941) at 145-159.

After the war had begun: W. Scott to G.B. McClellan, May 3, 1861, ORA Ser. 1, Vol. 51, Pt. 1 at 369-370. E.D. Townsend, Anecdotes of the Civil War in the United States (1883) at 55-56.

Immediately after the attack: A. Lincoln, Memorandum, Jul. 23, 1861, R.P. Basler, Collected Works of Abraham Lincoln, Vol. 4 at 457.

That the Federals had an advantage: Population of the United States in 1860, Eighth Census.

The 1860 census enumerated: Manufacture of the United States in 1860, Eighth Census.

While the military advantage implied: J.L. Harsh, Confederate Tide Rising at 178-184.

Several important crops grew: "Mr. David Chadwick on the Progress of Manchester from 1840 to 1860" in Report on the Thirty-First Meeting of the British Association for the Advancement of Science (1862) (Pt. 2) at 210. Agriculture of the United States in 1860, Eighth Census.

In theory, a blockade runner: R.C. Black III. Railroads of the Confederacy.

The Confederate coast—which: C.L. Symonds, The Civil War at Sea at 42.

In addition, the harbor needed: "The Harbor of Charleston", De Bow's Review, Vol. 26 (1859) at 222-223. "Harbor of Charleston, S.C.", ibid. at 698-701. I.S. Homans and I.S. Homans, Jr., eds., A Cyclopedia of Commerce and Commercial Navigation, Harper and Brothers, 1859 and 1860, at 290-291 and 1434.

The causation was circular: Report of the Secretary of the Treasury, Transmitting a Report from the Register of the Treasury for Commerce and Navigation for the Year Ending June 30, 1860 at 316-317, 350-351 and 522-523.

Principal and secondary harbors: S.R. Wise, Lifeline of the Confederacy at 149.

Just as a well-defended harbor provided: T. Bailey to G. Welles, Apr. 2, 1863, ORN Ser. 1, Vol. 17 at 40. G. Welles to S.P. Chase, Apr. 21, 1863, ibid. at 417.

Chapter 2. Evolving Tools of War

Steam made a vessel largely independent: E.W. Sloan III, Benjamin Franklin Isherwood Naval Engineer at 19 and 33-34. W.H. Roberts, Now for the Contest at 33-35.

The earliest steamboats used paddlewheels: S.R. Wise, Lifeline of the Confederacy.

The most profound effect of steam: T. Roosevelt, The Naval War of 1812 at 78-86.

Speed and the capacity for: Statistical Data of U.S. Ships and Confederate Ships, ORN Ser. 2, Vol 1 at 27-272.

These benefits did not come: The Boiler Explosion of the Martin Boiler on Board the U.S. "Double-Ender" Chenango, *passim*. T.A. Budd to S.F. Du Pont, Nov. 8, 1861, ORN Ser. 1, Vol. 13 at 280 (USS *Penguin* at Port Royal). USS *Penguin* (ship data), ORN Ser. I, Vol. 1 at 173 (screw steamer). F. Buchanan to S.R. Mallory, Mar. 27, 1862, ORN Ser. 1, Vol. 7 at 44 (CSS *Patrick Henry* at Hampton Roads). CSS *Patrick Henry* (ship data), ORN Ser. 2, Vol. 1 at 262 (side wheel steamer). H.S. Stellwagen to S.F. Du Pont, Jan. 31, 1863, ORN Ser. 1, Vol. 13 at 579 (USS *Mercedita* at Charleston). USS *Mercedita* (ship data), ORN Ser. 2, Vol. 1 at 141 (screw steamer). W.E. Le Roy to S.F. Du Pont, Jan. 31, 1864, ORN Ser. 1, Vol. 13 at 582 (USS *Keystone State* at Charleston). USS *Keystone State*, (ship data), ORN Ser. 2, Vol. 1 at 120 (side wheel steamer). F.A. Roe to M. Smith, May 5, 1864, ORN Ser. 1, Vol. 9 at 737 (USS *Sassacus* in North Carolina). USS *Sassacus* (ship data), ORN Ser. 2, Vol. 1 at 202 (screw steamer). C.L. Huntington to D.G. Farragut, Aug. 6, 1864, ORN Ser. 1, Vol. 21 at 479 (USS *Oneida*). USS *Oneida* (ship data), ORN Ser. 2, Vol. 1 at 165 (screw steamer). J.T. Seaver to D.G. Farragut, Aug. 6, 1864, ORN Ser. 1, Vol. 21 at 506-07 (USS *Philippi* at Mobile). USS *Philippi* (ship data), ORN Ser. 2, Vol. 1 at 17 (side wheel steamer). J.C. Beaumont to D.D. Porter, Dec. 27, 1864, ORN Ser. 1, Vol. 11 at 319 (USS *Mackinaw* at Wilmington). USS *Mackinaw* (ship data), ORN Ser. 2, Vol. 1 at 130 (side wheel steamer). J.M.B. Clitz to D.D. Porter, Dec. 27, 1864, ORN Ser. 1, Vol. 11 at 336 (USS *Osceola* at Wilmington). USS *Osceola* (ship data), ORN Ser. 2, Vol. 1 at 167 (side wheel steamer).

The increased destructiveness of naval: J.P. Baxter III, The Introduction of the Ironclad Warship at 209.

The pendulum swing toward defense: G.V. Fox to J.W. Grimes, Mar. 12, 1862, ORN Ser. 1, Vol. 7 at 98. S.C. Tucker, Arming the Fleet at 220. R.J. Schneller, Jr., "A State of War is a Most Unfavorable Period for Experiments", International Journal of Naval History, Dec. 2003, Vol. 2, No. 3 at 3 and 14-15. Survey of CSS Atlanta, Jun. 22, 1863, ORN Ser. 1, Vol. 14 at 275. Survey of Ordnance and Ordnance Stores aboard CSS Atlanta, Jun. 30, 1863, ibid. at 278. Report of the Secretary of the Navy, Nov. 5, 1864, ORN Ser. 2, Vol. 2 at 757.

Another approach was to use: D.G. Farragut to H.A. Wise, Jun. 11, 1864, ORN Ser. 1, Vol. 21 at 331-332. D.G. Farragut to G.V. Fox, Jun. 14, 1864, ORN Ser. 1, Vol. 21 at 335. R. Sherman to T.H. Stevens, Aug. 5, 1864, ORN Ser. 1, Vol. 21 at 498 (Monitor *Winnebago*). J.A. McDonald to G.H. Perkins, Aug. 7, 1864, ORN Ser. 1, Vol. 21 at 501 (Monitor *Chickasaw*).

The Federal Navy initially took: A.A. Harwood to J. Smith, Feb. 7, 1862, ORN Ser. 1, Vol. 6 at 604. L.M. Goldsborough to W.N. Jeffers, Mar. 16, 1862, ORN Ser. 1, Vol. 7 at 28. W.N. Jeffers to L.M. Goldsborough, Mar. 16, 1862, ORN Ser. 1, Vol. 7 at 28-29.

From time to time a story that is: Testimony of I. McDowell, Report of the Joint Committee on the Conduct of the War, Part 2 Bull Run, 37th Cong., 3rd Sess. at 36 and 38. E.D. Townsend, Anecdotes of the Civil War at 57.

The observation is relevant here: Dictionary of American Naval Fighting Ships, *Princeton I* (Screw Steamer), https://www.history.navy.mil/research/histories/ship-histories/danfs/p/princeton-i.html. W.C. Church, The Life of John Ericsson, Vol. 1 at 117-131. O. Thulesius, The Man who Made the Monitor at 56-66. R.J. Schneller, Jr., "A State of War is a Most Unfavorable Period for Experiments", International Journal of Naval History (Dec. 2003), Vol. 2, No. 3 (www.ijnhonline.org at 8). Ordnance Instruction for the United States Navy, 1866, Part 3 at 53. Testimony of I. Newton, Report of the Joint Committee on the Conduct of the War, Light Draft Monitors, 38th Cong., 2nd Sess. at 53.

Although the design of each: The information in this paragraph and the several that follow is from J.R. Weaver II, A Legacy in Brick and Stone at 7, 19-21, 24, 26-30, 32-33, 209 and 211. J.E. Kaufmann and H.W. Kaufmann, Fortress America at 143 and 208. C. Duffy, Fire & Stone at 10 and 24-32. E.R. Lewis, Seacoast Fortifications of the United States at 23-24, figures 5 and 6, 43-45 and 51, figures 24 and 24. W.B. Robinson, American Forts at 8 and 197-198.

The Confederates placed obstructions: Report of Q.A. Gillmore, Oct. 20. 1865, ORA Ser. 1, Vol. 6 at 148-165.

After the war, Lieutenant John A. Dahlgren: The information in this paragraph and the several that follow is from J.A. Dahlgren, Boat Armament of the U.S. Navy (1856).

Boat howitzers saw much use: General order of Flag-Officer Farragut, no date, ORN Ser. 1, Vol. 18 at 48. R.H. Wyman to T.T. Craven, Sep. 11, 1861, ORN Ser. 1, Vol. 4 at 667.

Communications at sea were: The information about signals and signal codes in this paragraph and the several that follow is from Code of Flotilla and Boat Squadron Signals for the United States Navy (1861).

In 1858 a British Navy officer: "Signals, Naval" in Encyclopaedia Britannica (1894) Vol. 22 at 55.

The flames of lamps and candles illuminated: G. Ripley and C.A. Dana, eds., The New American Cyclopaedia (1859), Vol. 6 at 629. "The Skating Carnival", *New York Times*, Dec. 19, 1860.

Calcium lights were not standard: Testimony of M. Lovell, Apr. 8, 1863, ORA Ser. 1, Vol. 6 at 562. E.W. Serrell to Q.A. Gillmore, Sep. 10, 1863, ORA Ser. 1, Vol. 28, Pt. 2 at 235. T.B. Brooks to Q.A. Gillmore, Sep. 27, 1863, ORA Ser. 1, Vol. 28, Pt. 2 at 267. R.S. Ripley to T. Jordan, ORA Ser. 1, Vol. 28, Pt. 2 at 386. W. Tennent, Jr. to D.B. Harris, Aug 13, 1863, ORA Ser. 1, Vol. 28, Pt. 2 at 510. Report of A. Rhett, ORA Ser. 1, Vol. 28, Pt. 2 at 578. Report of A. Rhett, ORA Ser. 1, Vol, 28. Pt. 1 at 578 (entry for Aug. 10, 1863). P.G.T. Beauregard to S. Cooper, no date, ORA Ser. 1, Vol. 28, Pt. 1 at 82. R.S. Ripley to T. Jordan, Aug. 21, 1863, ORA Ser. 1, Vol. 28, Pt. 1 at 386. W.H.C. Whiting to J.F. Gilmer, Dec. 23, 1864, ORA Ser. 1, Vol. 42, Pt. 3 at 1297. J.A. Dahlgren to G. Welles, Apr. 6, 1864, ORN Ser. 1, Vol. 15 at 394. J.W. Turner to Lt. Burbank, Feb. 26, 1865, ORA Ser. 1, Vol. 46, Pt. 2 at 710.

One curious omission is that the Federals: S.P. Quackenbush to G. Welles, Feb. 2, 1863, ORN Ser. 1, Vol.13 at 640.

Admiral S.P. Lee requested: S.P. Lee to G. Welles, Apr. 20, 1864, ORN Ser. 1, Vol. 9 at 664. G.V. Fox to D.D. Porter, Nov. 19, 1864, ORN Ser. 1, Vol. 11 at 78.

Ossabaw Sound was a large inlet: J.R. Goldsborough to S.F. Du Pont, Jul. 26, 1862, ORN Ser. 1, Vol. 13 at 217. Note from S.B. Gregory, no date, ORN Ser. 1, Vol. 13 at 193-194. S.F. Du Pont to G.V. Fox, Aug. 21, 1862, ORN Ser. 1, Vol. 13 at 269.

Four Federal gunboats made: H. Bryan to T. Jordan, Feb.1, 1863, ORA Ser. I, Vol. 14 at 212-223. J.L. Worden to S.F. Du Pont, Feb. 2, 1863, ORN Ser. 1, Vol. 13 at 628. Official Military Atlas of the Civil War at Plate 70, Map 2. Report of J. McCrady, Mar. 8, 1863, ORA Ser. 1, Vol. 14 at 222-223. C. Steedman to S.F. Du Pont, Jul. 29, 1862, ORN Ser. 1, Vol. 13 at 221. J.L. Davis to S.F. Du Pont, Nov. 19, 1862, ibid. at 454. J.L. Worden to S.F. Du Pont, Jan. 27, 1863, ibid. at 544-545. P.G.T. Beauregard to S. Cooper, Jan. 28, 1863, ibid. at 550. T.A. Stephens to J.L. Worden, Jan. 27, 1863, ibid. at 545-546.

The *Montauk* and the other Federal: J.L. Worden to S.F. Du Pont, Feb. 2, 1863, ORN Ser. 1, Vol. 13 at 628-629. H. Bryan to T. Jordan, Feb. 1, 1863, ibid. at 635. R.H. Anderson to G.A. Mercer, Feb. 2, 1863, ibid. at 635-637. T.A. Stephens to J.L. Worden, Feb. 2, 1863, ibid. at 631-632. R.B. Gordon, American Iron at 7. J.L. Worden to S.F. Du Pont, Feb. 2, 1863, ORN Ser. 1, Vol. 13 at 630-631. T.A. Stephens to J.L. Worden, Feb. 2, 1863, ibid. at 632.

On February 27, 1863, Admiral Du Pont: S.F. Du Pont to P. Drayton, Feb. 27, 1863, ORN Ser. 1, Vol. 13 at 694-695. J.L. Worden to S.F. Du Pont, Feb. 28, 1863, ibid. at 697-698. S.T. Browne, First Cruise of the Montauk at 54-56. A second accident occurred as: G.W. Anderson, Jr. to G.A. Mercer, Feb. 28, 1863, ibid. at 708-709. Report of Board of Survey, Mar. 5, 1863, ibid. at 707-708. T.A. Stephens to J.L. Worden, Feb. 28, 1863, ibid. at 700-704.

On March 3, 1863, the monitors *Passaic*: S.F. Du Pont, to G. Welles, Mar. 6, 1863, ORN Ser. 1, Vol. 13 at 716. P. Drayton to S.F. Du Pont, Mar. 4, 1863, ibid. at 717-718. P. Drayton to J.A. Dahlgren, Mar. 8, 1863, ibid. at 727. J.N. Miller to P. Drayton, Mar. 4, 1863, ibid. at 719-720 (USS *Passaic*). D. Ammen to P. Drayton, Mar. 3, 1863, ibid. at 721 (USS *Patapsco*). J. Downes to P. Drayton, Mar. 4, 1863, ibid. at 722 (USS *Nahant*).

A Confederate engineer at Fort McAllister: D.B. Harris to T. Jordan, Mar. 9, 1863, ORN Ser. 1, Vol. 13 at 730. J. McCrady to D.B. Harris and enclosure, Mar. 8, 1863, ORN Ser. 1, Vol. 13 at 731-734.

Chapter 3: Exceptional Commercial Circumstances

The blockade running during the Civil War: A Report from the Register of the Treasury of the Commerce and Navigation of the United States for the Year Ending June 30, 1861 at 513 and 516-517. Computed from Report of the Secretary of the Treasury Transmitting a Report from the Register of the Treasury, of Commerce and Navigation of the United States for the Year Ending June 30, 1861 at 502.

Chart 3-1: Captures of Coasting Vessels: Atlantic Coast: S.H. Stringham to G. Welles, May 14, 1861, ORN Ser. 1, Vol. 5 at 630-631 (schooners *Mary Willis*, *Delaware Farmer* and *Emily Ann*). G.J. Pendergrast to U.S. District Judge, May 26, 1861, ibid. at 673 (American schooner *Iris*). G.J. Pendergrast to U.S. District Judge, May 26, 1861, ibid. at 674 (Confederate schooner *Catherine*). G.J. Pendergrast to G. Welles, Jun. 8, 1861, ibid. at 709 (small sloops). Gulf Coast: T.A. Craven to G. Welles and enclosures, Jun. 22, 1861, ORN Ser. 1, Vol. 16 at 558-559 (sloop *President Filmore*). M. Smith to W. Mervine, Jun. 25, 1861, ibid. at 560 (schooner *Trois Freres*; schooner *Olive Branch*; schooner *Fanny*; schooner *Basilde*).

J. Alden to W. Mervine, Aug. 28, 1861, ibid. at 578 (schooner *Tom Hicks*. schooner *General T.J. Chambers*). M. Smith to W. Mervine, Aug. 8, 1861, ibid. at 615-616 (sloop *Charles Henry*). C. Price to W. Mervine, Sep. 25, 1861, ibid. at 687 (schooner *Cecilia*). C. Price to G. Welles, Oct. 18, 1861, ibid. at 694-695 (schooner *Zavala*). G.F. Emmons to G. Welles and enclosures, Jul. 6, 1862, ORN Ser. 1, Vol. 18 at 666-667 (schooner *Sarah*).

In addition to the coastal traffic: J. McNeil, Masters of the Shoals at 32. J.R. Soley, The Blockade and the Cruisers at 153.

Chart 3-2: Captures of Sailing Vessels: Atlantic Coast: W.W. McKean to G. Welles, May 12, 1861, ORN Ser. 1, Vol. 5 at 629 (ship *General Parkhill*). S.H. Stringham to G. Welles, May 14, 1861, ibid. at 631 (ship *Argo*). S.H. Stringham to G. Welles, May 17, 1861, ibid. at 639 (American bark *Star*). S.H. Stringham to G. Welles, May 25, 1861, ibid. at 669 (Confederate bark *Sophia*). G.J. Pendergrast to U.S. District Judge, May 30, 1861, ibid. at 683 (Confederate schooner *Lynchburg*). S.H. Stringham to G. Welles, Jun. 11, 1861, ibid. at 714 (Confederate brig *Hallie Jackson*). J.R. Goldsborough to G. Welles, Aug. 31, 1861, ibid. at 729 (Confederate ship *Amelia*). S.H. Stringham to G. Welles, Jun. 26, 1861, ibid. at 749 (Confederate bark *Sally Magee*). S.H. Stringham to G. Welles, Jul. 11, 1861, ibid. at 784 (Confederate brig *Amy Warwick*). **North Atlantic Coast:** S.H. Stringham to G. Welles, Jul. 20, 1861, ORN Ser. 1, Vol. 6 at 18 (Confederate schooner *Velasco*). S. Mercer to S.H. Stringham, Aug. 6, 1861, ibid. at 62 (British brigantine *Sarah Starr*). J.F. Armstrong to G. Welles, Aug. 24, 1862, ORN Ser. 11, Vol. 7 at 670 (British schooner *Mary Elizabeth*). D.L. Braine to G. Welles, Oct. 11, 1862, ORN Ser. 1, Vol. 8 at 128 (British schooner *Revere*). J.M.B. Clitz to S.P. Lee, Nov. 3, 1862, ibid. at 190 (British schooner *Pathfinder*). J. Trathen to G. Welles, Mar. 24, 1863, ibid. at 626 (British schooner *Mary Jane*). C.S. Boggs to S.P. Lee, May 2, 1863, ibid. at 837 (British schooner *Wanderer*). S.P. Lee to G. Welles, Nov. 7, 1863, ORN Ser. 1, Vol. 9 at 269 (British schooner *Herald*). E. English to W.H. Macomb, Jan. 9, 1865, ORN Ser. 1, Vol. 11 at 422 (schooner *Triumph*). **South Atlantic Coast:** H.Y. Perviance to G. Welles, Nov. 7, 1861, ORN Ser. 1, Vol. 12 at 331-332 (British schooner *Fanny Lee*). S.F. Du Pont to G. Welles, Apr. 3, 1862, ibid. at 369 (schooner *E.J. Waterman*). C. Steedman to S.F. Du Pont, Dec. 13, 1861, ibid. at 394 (schooner *Sarah and Carolina*). S.F. Du Pont to G. Welles, Jan. 7, 1862, ibid. at 428 (British schooner *Prince of Wales*; British schooner *Island Belle*). S.F. Du Pont to G. Welles, May 3, 1862, ibid. at 708 (British schooner *British Empire*). P.G. Watmough to T. Turner, Mar. 31, 1863, ORN Ser. 1, Vol. 13 at 800-801 (British schooner *Antelope*). E.R. Colhoun to S.F. Du Pont, Apr. 20, 1863, ORN Ser. 1, Vol. 14 at 151 (British schooner *Minnie*). F.H. Baker to J.A. Dahlgren, Jan. 3, 1864, ORN Ser. 1, Vol. 15 at 219-220 (British schooner *Sylvanus*). J.F. Green to J.A. Dahlgren and enclosure, Jul. 9, 1864, ibid. at 563-564 (British schooner *Pocahontas*). S. Beaumont to G. Welles, Dec. 3, 1864, ORN Ser. 1, Vol. 11 at 125 (British schooner *Mary*). S.H. Bryant to J.A. Dahlgren, Jan. 28, 1865, and J.C. Chaplin to J.A. Dahlgren, Feb. 3, 1865, ORN Ser. 1, Vol. 16 at 201 (Confederate schooner *Coquette*). **Gulf Coast:** M. Smith to W. Mervine, Jun. 25, 1861, ORN Ser. 1, Vol. 16 at 560 (Mexican schooner *Brilliante*). J. Alden to W. Mervine, Sep. 13, 1861, ibid. at 665-666 (Mexican schooner *Soledad Cos*). J. Alden to W.W. McKean, Oct 17, 1861, ibid. at 732 (British schooner *Edward Barnard*). J. Alden to W.W. McKean, Oct. 4, 1861, ibid. at 738-739 (British schooner *Ezilda*; British schooner *Joseph H. Toone*). A. Read to G. Welles, Dec. 30, 1861, ORN Ser. 1, Vol. 17 at 21 (schooner *Gipsey*). W.M. Walker to G. Welles, Feb. 3, 1862, ibid. at 88 (schooner *Major Barbour*). **East Gulf Coast:** D. Cate to G. Welles, Mar. 7, 1862, ORN Ser. 1,

Vol. 17 at 184 (Confederate schooner *Anna Belle*). J.L. Lardner to G. Welles, Sep. 29, 1862, ibid. at 314-315 (British schooner *Isabel*). W.C. Rogers to G. Welles, Mar. 13, 1863, ibid. at 384-385 (British schooner *Surprise*). J.S. Williams to C.E. Fleming, Oct. 30, 1863, ibid. at 581 (British schooner *Meteor*). F. Burgess to G. Welles, Feb. 1, 1864, ibid. at 641 (British sloop *Racer*). L.N. Stodder to G. Welles, Nov. 16, 1864, ibid. at 778 (Confederate schooner *Badger*). W. Gibson to G. Welles, Feb. 15, 1865, ibid. at 811 (schooner *Delia*). West Gulf Coast: P. Crosby to G. Welles, Mar. 9, 1862, ORN Ser. 1, Vol. 18 at 56 (British schooner *Cora*). G.F. Emmons to G. Welles and enclosures, Jul. 6, 1862, ibid. at 666-667 (Confederate sloop *Elizabeth*). J.R. Goldsborough to G. Welles, Mar. 25, 1863, ORN Ser. 1, Vol. 20 at 99 (schooner *Clara*). W.H. Dana to G. Welles, Aug. 22, 1863, ibid. at 475-476 (Swiss schooner *Wave*). R. Tarr to G. Welles, Feb. 12, 1864, ORN Ser. 1, Vol. 21 at 79 (British schooner *Louisa*). D.G. Farragut to G. Welles, Oct. 19, 1864, ibid. at 672 (British schooner *Annie Virden*). A.G. Clary to G. Welles, Jan. 14, 1865, ORN Ser. 1, Vol. 22 at 11 (Confederate schooner *Josephine*). J.A. Johnston to G. Welles, Apr. 22, 1865, ibid. at 136-137 (British schooner *Chaos*).

An article in the *Nassau Guardian*: "Intercourse of the British at Nassau with the Rebel States", *New York Times*, Apr. 28, 1862.

Most of the vessels that were captured: J.R. Soley, The Blockaders and the Cruisers at 44. S.R. Wise, Lifeline of the Confederacy at 221.

A report submitted shortly: R.W. Rawson, Report on the Bahamas for the Year 1864 at 54.

Cotton was compressed into bales: "The Cotton Crop of 1859-60", Merchant's Magazine and Commercial Review, Vol. 47 (1862) at 358. W. Watson, The Adventures of a Blockade Runner at 260.

Another early phenomenon: S.R. Wise, Lifeline of the Confederacy at 66, 67, 71 and 323.

The experience of the *Gladiator* set: S.R. Wise, Lifeline of the Confederacy at 323.

Dividing the traffic between Europe: H.S. Stellwagen to W.W. McKean, May 3, 1862, ORN Ser. 1, Vol. 17 at 220-221.

Not all the cargoes carried: L.H. Johnson, "Commerce Between Northeastern Ports and the Confederacy, 1861-1865", Journal of American History, Vol. 54, No. 1 (1967) at 30-42. E.M. Coulter, "Commercial Intercourse with the Confederacy in the Mississippi Valley", Mississippi Valley Historical Review, Vol. 5, No. 4 (1919) at 377-395. A.S. Roberts, "Federal Government and Confederate Cotton", American Historical Review, Vol. 32, No. 2 (1927) at 262-275. A.F. Smith, Starving the South at 15-16. P. Leigh, Trading with the Enemy.

The profits to be made from trading: L.H. Johnson III, "Blockade or Trade Monopoly?", Virginia Magazine of History and Biography, Vol. 93, No. 1 (1985) at 54-78. J.H. Parks, "A Confederate Trade Center Under Federal Occupation", Journal of Southern History, Vol. 7, No. 3 (1941) at 289-314. P. Leigh, Trading with the Enemy.

Northern goods and provisions: P. Leigh, Trading with the Enemy. F.I. Wilson, Sketches of Nassau at 14.

A later phase of blockade running: J. McNeil, Masters of the Shoals at 46-48. J.R. Soley, The Blockade and the Cruisers at 156-157. S.R. Wise, Lifeline of the Confederacy at 112-113. F.B.C. Bradlee, Blockade Running During the Civil War and the Effect of Land and Water Transportation on the Confederacy at 33-34.

To an observer at the surface of the ocean: J.B. Hattendorf, "The Royal Navy During the War of the French Revolution and the Napoleonic War" in D. King, A Sea of Word at 5. J.R. Soley, The Blockade and the Cruisers at 156-157. J. Wilkinson, The Narrative of a Blockade-Runner at 131-132.

Another blockade-running captain: W. Watson, Adventures of a Blockade-Runner at 160, 170 and 290-304.

The blockade runners took: J. McNeil, Masters of the Shoals at 46-48. S.R. Wise, Lifeline of the Confederacy at 110. J. Wilkinson, The Narrative of a Blockade-Runner at 153-155.

A sailing ship relied primarily upon: W. Watson, Adventures of a Blockade-Runner at 168 and 257.

Steamships used stealth when possible: J. Wilkinson, The Narrative of a Blockade-Runner at 156. W. Watson, Adventures of a Blockade-Runner at 44-45 and 309.

Although the blockade runner at sea: G. Welles to S.F. Du Pont, Nov. 7, 1861 and enclosure, ORN Ser. 1, Vol. 12 at 329-330. R.M. Browning Jr., Success is All That Was Expected at 287. J. Wilkinson, The Narrative of a Blockade-Runner at 155-156.

Blockade runners also received assistance: E. Lonn, Salt as a Factor in the Confederacy. J.R. Soley, The Blockade and the Cruisers at 158-161. W.L. Powell, Signals for Running Blockade, no date, ORN Ser. 1, Vol. 16 at 729-730.

At times blockade runners shared: G. Welles to S.F. Du Pont, Nov. 7, 1861 and enclosure, ORN Ser. 1, Vol. 12 at 329-330. R.M. Browning, Jr., Success is All That Was Expected at 287. J. Wilkinson, The Narrative of a Blockade-Runner at 155-156. J.R. Soley, The Blockade and the Cruisers at 136-139. S.R. Wise, Lifeline of the Confederacy at 62 and 82-83.

From the outset, blockade running: J.R. Soley, The Blockade and the Cruisers at 116-118.

Yet the Confederates also impeded blockade: F.L. Owsley, King Cotton Diplomacy at 24-43.

Even before the end of the embargo: R.C. Todd, Confederate Finance at 44-51.

By early 1863 the Confederates were becoming: S.R. Wise, Lifeline of the Confederacy at 73. T. Jordan to L. Heyliger and enclosure, May 6, 1863, ORA Ser. 1, Vol. 14 at 926-927. J.A. Seddon to P.G.T. Beauregard, Aug. 11, 1863, ORA Ser. 4, Vol. 2 at 714-715.

Just before the end of 1862: S.R. Wise, Lifeline of the Confederacy at 94-106 and 124. F.E. Vandiver, Ploughshares into Swords at 84-104.

Colin J. McRae, the Confederate fiscal agent: S.R. Wise, Lifeline of the Confederacy at 142-143 and 147-150. J. McRae to J.A. Seddon and enclosure, ORA Ser. 4 Vol.3 at 525-530. C.S. Davis, Colin J. McRae.

At the time, the Confederate government: S.R. Wise, Lifeline of the Confederacy at 147-150 and Appendix 22.

The Confederates attempted to exercise: AN ACT to prohibit the importation of luxuries, Feb. 6, 1864 ORA Ser. 4, Vol. 3 at 78-80.

The second law prohibited the export: A BILL to impose regulations upon the foreign commerce of the Confederate States, Feb. 6, 1864, ORA Ser. 4, Vol. 3 at 80-82.

By the summer of 1864 the three principal: "The Seaports of the Confederacy", *New York Times*, Mar. 9. 1865.

Chapter 4: Planning the Blockade and Blockading Tactics

Although President Lincoln proclaimed: "Proclamation of a Blockade", Apr. 19, 1861, R.P. Basler, ed., Collected Works of Abraham Lincoln, Vol. 4 at 338-339. Proclamation of Blockade, Apr. 27, 1861, ibid. at 346-347. Cooperation of the Navy in the Relief of Fort Pickens, April 12 and 17, 1861, ORN Ser. 1, Vol. 4 at 107-138. J.H. Strong to G. Welles, Mar. 30, 1861, ORN Ser. 1, Vol. 4 at 103. J.H. Strong to G. Welles, Apr. 12, 1861, ibid. at 106-107. J.H. Strong to G. Welles, Apr. 19, 1861, ibid. at 139-140. Reports and correspondence relative to the destruction and abandonment of the Norfolk navy yard, April 20, 1861, ibid. at 272-313. G. Welles, to G.S. Blake, Apr. 27, 1861, ibid. at 340. S.F. Du Pont to G. Welles and enclosures, Sep. 23, 1861, ibid. at 394-398. G. Welles to T.S. Fillebrown, Apr. 20, 1861, ibid. at 415. C.F. Smith to G. Welles, Apr. 21, 1861, ibid. at 416-417. G. Welles to F. Buchanan, Apr. 21, 1981, ibid. at 417. C. Vanderbilt to G. Welles, Apr. 16, 1861, ORN Ser. 1, Vol. 1 at 8. Petition of New York merchants and bankers to the Secretary of the Treasury that means of defense be furnished the California steamers, Apr. 17, 1861, ibid. at 8. J.D. Jones to G. Welles and enclosures, Apr. 18, 1861, ibid. at 9.

At the start of President Lincoln's term: Report of the Secretary of the Navy, Jul. 4, 1861, Congressional Globe App., 37th Cong. 1st Sess. at 7-8. Report of the Secretary of the Navy, Dec. 2, 1861, Congressional Globe App., 37th Cong. 2nd Sess. at 20.

The president's proclamations in April 1861: "Proclamation of a Blockade", Apr. 19, 1861, R.P. Basler, ed., Collected Works of Abraham Lincoln, Vol. 4 at 338-339. "Proclamation of Blockade", Apr. 27, 1861, ibid. at 346-347. S.R. Wise, Lifeline of the Confederacy at 25. J.R. Soley, The Blockade and the Cruisers at 35 and 84. J.W. Livingston to S.H. Stringham, Aug. 15, 1861, ORN Ser. 1, Vol. 6 at 86.

Initially Secretary of the Navy Gideon: G. Welles to S.H. Stringham (2 letters), May 1, 1861, ORN Ser. 1, Vol. 5 at 619-620 and 621-622. G. Welles to W. Mervine, May 7, 1861, ORN Ser. 1, Vol. 16 at 519-520. W.H. Roberts, Now for the Contest at 11.

In June 1861 Secretary Welles assembled: G. Welles to J.G. Barnard, Jun. 26, 1861, ORN Ser. 1, Vol. 12 at 195. Report of the Superintendent of the Coast Survey (1861) at 2 and 87. J. Cloud, The U.S. Coast Survey in the Civil War at 18-21.

From July through September 1861: Conference Report (Atlantic), Jul. 5, 1861, ORN Ser. 1, Vol. 12 at 195-198. Conference Report (Atlantic), Jul. 13, 1861, ORA Ser. 1, Vol. 53 at 67-73. Conference Report (Atlantic), Jul. 16, 1861, ORN Ser. 1, Vol. 12 at 198-201. Conference Report (Atlantic), Jul. 26, 1861, ibid. at 201-206.

The conference's final reports focused: Conference Report (Gulf of Mexico), Aug. 9, 1861, ORN Ser. 1, Vol. 16 at 618-630. Conference Report (Gulf of Mexico), Sep. 3, 1861, ibid. at 651-655. Conference Report (Gulf of Mexico), Sep. 19, 1861, ibid. at 680-681.

The conference also recommended: Conference Report (Atlantic), Jul. 26, 1861, ORN Ser. 1, Vol. 12 at 206.

Even as the conference completed: G. Welles to S.F. Du Pont, Aug. 3, 1861, ORN Ser. 1, Vol. 12 at 207. G. Welles to S.F. Du Pont, Sep. 18, 1861, ibid. at 208.

With only a small fleet at the start: R.M Browning, Jr., Success Is All That Was Expected at 52-54. "Doomed Cities", *New York Times*, Nov. 27, 1861. R.E. Lee to J.P. Benjamin, Dec. 20, 1861, ORN Ser. 1, Vol. 12 at 423 and ORA Ser. 1, Vol. 6 at 42-43. Journal of R. Semmes, Jan. 27, 1862, ORN Ser. 1, Vol. 1 at 739. "American Topics: Intervention and the Stone Blockade", *New York Times*, Feb. 1, 1862. "The Destruction of Charleston Harbor", ibid., Mar. 5, 1862 (reporting remarks made Feb. 14, 1862). "Stone Blockades in English History", ibid., Feb. 2, 1862. "Mr. Seward's Explanation of the Stone Blockade", ibid., Feb. 13, 1862.

The turmoil raised over the stone: A.D. Bache to S.F. Du Pont, Sep. 4, 1861, ORN Ser. 1, Vol. 12 at 207. C.H. Davis to S.F. Du Pont, Dec. 21, 1861, ORN Ser. 1, Vol. 12 at 422. F.P. McKibben, "The Stone Fleet of 1861", New England Magazine, Jun. 1898, Vol. 24, Issue 4 at 488.

Just as the neutral powers feared: R. Werden to L.M. Goldsborough, Oct. 13, 1861, ORN Ser. 1, Vol. 6 at 315-316. L.M. Goldsborough to R. Werden, Oct. 14, 1861, ibid. at 318-319. R. Werden to L.M. Goldsborough, Oct 20, 1861, ibid. at 344-345. R. Werden to L.M. Goldsborough, Nov 2, 1861, ibid. at 377-378. L.M. Goldsborough to R. Werden, Nov. 6, 1861, ibid. at 410. R. Werden to L.M. Goldsborough, Nov. 8, 1861, ibid. at 414-415. R. Werden to L.M. Goldsborough, Nov. 17, 1861, ibid. at 428-430. L.M. Goldsborough to G. Welles, Nov. 7, 1861, ORN Ser. 1, Vol. 6 at 412. Unisys Weather: Atlantic Tropical Storm Tracking by Year, http://weather.unisys.com/hurricane/atlantic/. L.M. Goldsborough, Nov. 20, 1861, ORN Ser. 1, Vol. 6 at 430.

Ideas of how the blockade could be: G. Welles to S.H. Stringham, Jun. 5, 1861, ORN Ser. 1, Vol. 5 at 702. R.M. Browning, Jr., From Cape Charles to Cape Fear at 7. J.R. Soley, The Blockade and the Cruisers at 43. S.H. Stringham to G. Welles, May 24, 1861, ORN, Ser. 1, Vol. 5 at 664. W.W. Mervine, Oct. 16, 1861, ORN Ser. 1, Vol. 16 at 731.

On November 26, 1861, Captain James L. Lardner: J.L. Lardner to S.F. Du Pont, Nov. 26, 1861, ORN Ser. 1, Vol. 12 at 361. S.F. Du Pont to G. Welles, Dec. 4, 1861, ibid. at 380. F.W. Seward to G. Welles, Nov. 6, 1861, ibid. at 329-330. F.W. Seward to G. Welles, Nov. 21, 1861, ibid. at 355. G. Welles to S.F. Du Pont, Nov. 23, 1861, ibid. at 360. F.W. Seward to G. Welles, Dec. 12, 1861 and enclosure, ibid. at 398-399. G. Welles to S.F. Du Pont, Dec. 14, 1861, ibid. at 401. S. Whiting to F.W. Seward, Feb. 12, 1862, ibid. at 562.

This sanguine appraisal was dashed: "Affairs at Nassau: Important Information Regarding the Movements of Rebel Vessel - Running the Blockade, &c.", *New York Times*, Feb. 25, 1862. G. Welles to S.F. Du Pont, Mar. 28, 1862, ORN Ser. 1, Vol. 12 at 671. G. Welles to S.F. Du Pont and enclosure, Apr. 5, 1862, ibid. at 719-720.

The experience of the South Atlantic: E.M. Yard to S.F. Du Pont, Nov. 15, 1861, ORN Ser. 1, Vol. 12 at 346. T.A. Budd to S.F. Du Pont, Nov. 26, 1861, ibid. at 361. C. Steedman to S.F. Du Pont, Dec. 13, 1861, ibid. at 394. E.G. Parrott to S.F. Du Pont, Jan. 2, 1862, ibid. at 445. I.B. Baxter to S.F. Du Pont, Mar. 28, 1862, ibid. at 479-481. C. Steedman to S.F. Du Pont, Feb. 25, 1862, ibid. at 563. I.B. Baxter to S.F. Du Pont, Mar. 12, 1862, ibid. at 627. S.F. Du Pont to G. Welles, May 1, 1862, ibid. at 738. J.H. Upshur to S.F. Du Pont, Apr. 27, 1862, ibid. at 778-779. N. Collins to G. Welles, May 10, 1862, ibid. at 809. J.R. Beers to S.F. Du Pont, Apr. 22, 1862, ibid. at 760-761.

Flag Officer Du Pont wrote an upbeat response: S.F. Du Pont to G. Welles, Apr. 23, 1862, ORN Ser. 1, Vol. 12 at 771-773. S.F. Du Pont to C.H. Davis, Apr. 24, 1862, J.D. Hayes, ed., Samuel Francis Du Pont: A Selection From His Civil War Letters, Vol. 2 at 19 and n.1. G. Welles to J.P. Hale, May 9, 1862, ORN Ser. 1, Vol. 13 at 8-9.

Flag Officer Du Pont was more candid: S.F. Du Pont to G. Welles, Apr. 27, 1862, ORN Ser. 1, Vol. 12 at 782-783.

The combination of darkness and steam: S.F. Du Pont to S. Du Pont, Apr. 29, 1862, D.A. Hayes, Samuel Francis Du Pont: A Selection From His Civil War Letters Vol. 2 at 21.

With Captain Lardner reassigned: Chart, Charleston Harbor and neighborhood Showing Positions of U.S. Blockading Squadron under S.F. Du Pont Flag Officer, May 11, 1862, ORN Ser. 1. Vol. 12 at 816.

Commander Marchand rearranged: J.B. Marchand to S.F. Du Pont, Jun. 25, 1862, ORN Ser. 1, Vol. 13 at 138-139.

The success of the blockaders off: E.G. Parrott to S.F. Du Pont, May 11, 1862, ORN Ser. 1, Vol. 12 at 810. J.R.M. Mullany to S.F. Du Pont, May 24, 1862, ORN Ser. 1, Vol. 13 at 29.

At five on the morning of May 26: J. Downes to G. Welles, May 26, 1862, ibid. at 39.

At 4:45 on the morning of May 27: J.R.M. Mullany to G. Welles, May 27, 1862, ibid. at 47.

All three captures resulted from: U.S. Naval Observatory, Complete Sun and Moon Data for One Day, http://aa.usno.navy.mil/data/docs/RS_OneDay.php. S.R. Wise, Lifeline of the Confederacy at 292, 315 and 321-322.

The events as the next new Moon approached: W.E. le Roy to S.F. Du Pont, Jun. 20, 1862, ORN Ser. 1, Vol. 13 at 120. J.B. Marchand to S.F. Du Pont, Jun. 20, 1862, ibid. at 120-121. E. Lanier to G. Welles, Jun. 26, 1862, ibid. at 121. S.F. Du Pont to G. Welles, Jun. 27, 1862, ORN Ser. 1, Vol. 13 at 134-135. J.B. Marchand to S.F. Du Pont, Jun. 24, 1862, ibid. at 135-136. W.E. Le Roy to S.F. Du Pont, Jun. 25, 1862, ibid. at 136-137.

In September 1862 the North Atlantic: S.P. Lee to G.H. Scott, Sep. 21, 1862, ORN Ser. 1, Vol. 8 at 80. S.P. Lee to G. Welles and enclosure, Sep. 22, 1862, ORN Ser. 1, Vol. 8 at 81-82.

The number of vessels available to blockaders: General Orders No. 18, Oct. 22, 1864, ORN Ser. 1, Vol. 10 at 580.

The increased resources and revised tactics: C.L. Symonds, ed., Charleston Blockade at 140-141. Circular order of J.B. Marchand, Jun. 21, 1862, ORN Ser. 1, Vol. 13 at 131. J.B. Lippincott & Co., pubs., An American Dictionary of the English Language (1857) at 733. Brewer and Tileston, pubs., A Comprehensive Dictionary of the English Language (1860) at 327. Circular order of J.B. Marchand, Jul. 5, 1862, ORN Ser. 1, Vol. 13 at 169. General Instruction Regarding Signals, Aug. 25, 1862, ORN Ser. 1, Vol. 13 at 280-81.

The need to confirm friendly identity was: General instructions regarding signals, Aug. 25, 1862, ORN Ser. 1, Vol. 13 at 280-281. Circular order of S.W. Godon, Sep. 17, 1862, ORN Ser. 1, Vol. 13

at 328-329. Memorandum of T.A. Jenkins, Nov. 12, 1862, ORN Ser. 1, Vol. 19 at 343. General Orders No. 18, Oct. 22, 1864, ORN Ser. 1, Vol. 10 at 580-581.

Although blockade runners ventured from: E.G. Parrott to S.F. Du Pont, May 11, 1862, ORN Ser. 1, Vol. 12 at 810.

As the blockaders developed the tactics: H.H. Savage to B.D. Sands, Feb. 20. 1863, ORN Ser. 1, Vol. 8 at 561. H.H. Savage to B.F. Sands, Feb. 23, 1863, ibid. at 562. H.H. Savage to B.F. Sands, Feb. 25, 1863, ibid. at 568. S.P. Lee to B.F. Sands, Mar. 7, 1863, ORN Ser. 1, Vol. 8 at 589. General Orders No. 18, Oct. 22, 1864, ORN Ser. 1, Vol. 10 at 579-583.

The various elements of the improved: S.F. Du Pont to J.J. Almy, Dec. 12, 1862, ORN Ser. 1, Vol. 13 at 478. S.F. Du Pont to S.P. Quackenbush, Dec. 20, 1862, ORN Ser. 1, Vol. 13 at 48. W.R. Taylor to S.F. Du Pont and enclosures, Jan. 27, 1863, ORN Ser. 1, Vol. 13 at 538-539. S.P. Quackenbush to S.F. Du Pont, Feb. 23, 1863, ibid. at 539-540. S.F. Du Pont to G. Welles, Jan. 31, 1863, ORN Ser. 1, Vol. 13 at 551-552. P.H. Silverstone, Civil War Navies at 33-34 and 61.

Although the tactics that improved: P.H. Silverstone, Civil War Navies at 3-12. Instruction of S.P. Lee, Apr. 13, 1863, ORN Ser. 1, Vol. 8 at 802. Instructions of S.P. Lee, Apr. 21, 1863, ORN Ser. 1, Vol. 8 at 803.

In July 1863 the Navy Department replaced: D.D. Porter, The Naval History of the Civil War at 662-663. E.g., J.A. Dahlgren to G. Welles, Jul. 21, 1863, ORN Ser. 1, Vol. 14 at 374. Order of J.A. Dahlgren, Dec. 3, 1863, ORN Ser. 1, Vol. 15 at 148-149. Order of J.A. Dahlgren, Jul. 12, 1863, ORN Ser. 1, Vol. 14 at 338. J.A. Dahlgren to J.F. Green, Jul. 19, 1863, ibid. at 373. J.A. Dahlgren to G. Welles, Jul. 21, 1863, ORN Ser. 1, Vol. 14 at 374.

In December 1863 Admiral Dahlgren: Order Regarding Picket Duty, Dec. 3, 1863, ORN Ser. 1, Vol. 15 at 148-149. Order, Jan. 7, 1864, ORN Ser. 1, Vol. 15 at 226-227. W. Gibson to J.A. Dahlgren, Jan. 8, 1864, ibid. at 234. J.A. Dahlgren to G. Welles, Apr. 6, 1864, ORN Ser. 1, Vol. 15 at 394.

The combination of the presence of: S.R. Wise, Lifeline of the Confederacy at 253 and 257. W.H.C. Whiting to J.A. Seddon, Sep. 8, 1863, ORA Ser. 1, Vol. 29, Pt. 2 at 703-704. J.D. Bulloch to J. Low, Jan. 8, 1865, ORN Ser. 2, Vol. 2 at 788. J.F. Green to J.A. Dahlgren, Sep. 11, 1864, ORN Ser. 1, Vol. 15 at 670. General Order of J.A. Dahlgren, Sep. 16, 1864, ibid. at 680-682. Order of J.A. Dahlgren, Sep. 23, 1864, ibid. at 685-688.

Chapter 5: Mounting and Maintaining the Blockade

As the number of vessels increased: J.R. Soley, The Blockade and the Cruisers at 16 and 20. S.F. Du Pont to G. Welles, Apr. 27, 1862, ORN Ser. 1, Vol. 12 at 782. D.G. Farragut to G. Welles, Mar. 27, 1862, ORN Ser. 1, Vol. 18 at 86.

At the time of President Lincoln's inauguration: Report of the Secretary of the Navy, Jul. 4, 1861, Congressional Globe App., 37th Cong. 1st Sess. at 7-8. J.R. Soley, The Blockade and the Cruisers at 85 and 121. Report of the Secretary of the Navy, Dec. 2, 1861, Congressional Globe App., 37th Cong. 2nd Sess. at 20-21.

The size of the fleet at different times: Report of the Secretary of the Navy, Dec. 1, 1862, Congressional Globe App., 37th Cong. 3rd Sess. at 11 *et seq*. Report of the Secretary of the Navy, Dec. 7, 1863, Congressional Globe App., 38th Cong. 1st Sess. at 13 *et seq*. Report of the Secretary of the Navy, Dec. 5, 1864, Congressional Globe App., 38th Cong. 1st Sess. at 4 *et seq*. Report of the Secretary of the Navy, Dec. 2, 1861, Congressional Globe App., 37th Cong. 2nd Sess. at 20-21. R.M Browning, Jr., From Cape Charles to Cape Fear at 146. S.R. Wise, Lifeline of the Confederacy at 160-161.

The rapid expansion of the Navy's: C. Symonds, Lincoln and His Admirals at 57-59.

As a part of the attempt to destroy: "Reports and correspondence relative to the destruction and abandonment of the Norfolk navy yard, April 20, 1861", ORN Ser. 1, Vol. 4 at 272-313. W.C. Davis, Duel Between the First Ironclads at 7, 22 and 26-27. W.H. Roberts, Civil War Ironclads at 11.

The Federal Navy's ship constructors: R.G. Albion, Makers of Naval Policy at 196. "Report on Iron-Clad Vessels" in Message of the President of the United States to the Two Houses of Congress at the Commencement of the Second Session of the Thirty-Seventh Congress at 748.

Monitor's cannon could not fire: S. Tucker, Blue and Gray Navies at 171. W.C. Church, The Life of John Ericsson, Vol. 1 at 262-263 and 280-281.

A conventional warship provided: P.H. Silverstone, Civil War Navies at 4 and 15-16. W.C. Davis, Duel of the First Ironclads at 45.

On March 8, 1862, the CSS *Virginia*: J. Marston to G. Welles, Mar. 9, 1861, ORN Ser. 1, Vol. 7 at 8.

The news from Hampton Roads sent: Diary of Gideon Welles, Vol. 1 at 60-63.

The *Monitor* arrived at Hampton Roads: J. Marston to G. Welles, Mar. 9, 1861, ORN Ser. 1, Vol. 7 at 8. A. Pendergrast to G. Welles, Mar. 19, 1862, ibid. at 24. W. Radford to G. Welles, Mar. 10, 1865, enclosure, ibid. at 22. W.M. Wood to L. M. Goldsborough, Mar. 12, 1862, enclosure, ibid. at 17. W.C. Davis, Duel of the First Ironclads at 131-137.

The *Virginia* had sustained some: W.C. Davis, Duel of the First Ironclads at 134-137. F. Buchanan to S.R. Mallory, Mar. 27, 1862, ORN Ser. 1, Vol. 7 at 46 and 49. J.E. Tennent, The Story of the Guns (1864) at 286-292. R.J. Schneller, Jr., "A State of War is a Most

Unfavorable Period for Experiments", International Journal of Naval History (Dec. 2003), Vol. 2, No. 3 (www.ijnhonline.org at 8). A.A. Harwood to J. Smith, Feb. 7, 1862, ORN Ser. 1, Vol. 6 at 604. A.C. Stimers to J. Smith. Mar. 17, 1862 ORN Ser. 1, Vol. 7 at 27.

Had the *Monitor* not arrived: W.C. Davis, The Duel Between the First Ironclads at 47, 107 and 113-114. W.H. Parker, Recollections of a Naval Officer 1841-1865 at 288-289. R.M. Browning, Jr., From Cape Charles to Cape Fear at 46.

Before the war ended, the Federals: P.H. Silverstone, Civil War Navies at 4-10.

Twenty other vessels were smaller: Testimony of Isaac Newton, Dec. 24, 1864, Report of the Joint Committee on the Conduct of the War (Light-Draught Monitors) at 48-53. P.H. Silverstone, Civil War Navies at 9-10.

The Federals also built numerous: P.H. Silverstone, Civil War Navies at 4-12.

To compensate for the limited reserve: H.L. Wait, "The Blockade of the Confederacy", The Century, Oct. 1898, Vol. 56, Issue 6 at 927. J.A. Dahlgren to G. Welles, Dec. 6, 1863, ORN Ser. 1, Vol. 15 at 162. J.A. Dahlgren to G. Welles, Dec. 8, 1863, ORN Ser. 1, Vol. 15 at 163. Finding of the Court of Enquiry, ORN Ser. 1, Vol. 15 at 167. Memorandum of J.A. Dahlgren, Jan. 6, 1864, ORN Ser. 1, Vol. 15 at 168-169.

As the sectional crisis developed: C.B. Boynton, The History of the Navy During the Rebellion, Vol. 1 at 30-35. C.L. Symonds, Lincoln and His Admirals. J.R. Soley, The Blockade and the Cruisers at 9. G. Welles to J.A. Dahlgren, Oct. 14, 1863, ORN Ser. 1, Vol. 15 at 36-37. J.A. Dahlgren to G. Welles, Oct. 30, 1863, ibid. at 92-93.

Moreover, as the war progressed: G. Welles to J.A. Dahlgren, Oct. 14, 1863, ORN Ser. 1, Vol. 15 at 36-37. J.A. Dahlgren to G. Welles, Oct. 30, 1863, ORN ibid. at 92-93.

Along with ships and officers: J.R. Soley, The Blockade and the Cruisers at 9-10. W.H. Roberts, Now for the Contest at 143 and 205 n. 20. Report of the Secretary of the Navy, Dec. 7, 1863 Congressional Globe App., 38th Cong. 1st Sess. at 19.

Manpower shortages prompted the Navy: H. Aptheker, "The Negro in the Union Navy", The Journal of Negro History, Vol. 32, No. 2 (Apr. 1947) at 170-76. G. Welles to L.M. Goldsborough, Sep. 25, 1861, ORN Ser. 1, Vol 6 at 252. G. Welles to D.G. Farragut, Jun. 27, 1863, ORN Ser. 1, Vol. 20 at 322. G. Welles to H.H. Bell, Dec. 11, 1863, ibid. at 717.

The adequacy of supply was itself: G. Welles to W. Mervine, July 8, 1861, ORN Ser. 1, Vol. 27 at 356-357. G. Welles to S.H. Stringham, Jul. 16, 1861, ibid. at 357. W. Mervine to G. Welles, Sep. 19, 1861, ibid. at 365. G. V. Fox to H. Hastings, Nov. 26, 1861, ibid. at 384. G. Welles, to S.P. Chase, Jan. 4, 1862, ibid. at 396-397. G. Welles to S.P. Chase, May 1862, ibid. at 432. Instructions for the Government of Inspectors-in-Charge of Stores, Paymasters, and Assistant Paymasters (1865). Ordnance Instructions for the United States Navy (1860).

Supply was difficult from the first: Instructions for the Government of Inspectors-in-Charge of Stores, Paymasters, and Assistant Paymasters (1865). Ordnance Instructions for the United States Navy (1860).

In addition to the supplies provided: Regulations of the Government of Sutlers on board Supply Steamers of the Navy, Mar. 24, 1863, General Orders and Circulars issued by the Navy Department from 1863 to 1887 at 505-506. W.B. Eaton to G. Welles, Aug. 5, 1864, ORN Ser. 1, Vol. 27 at 605. E. Barnes and J.A. Barnes eds.,

Naval Surgeon at 73 and 123. A.A. Harwood to M. Meigs, Feb. 23, 1863, ORN Ser. 1, Vol. 5 at 233. S.P. Lee to G. Welles, Mar. 25 1863, ORN Ser. 1, Vol. 8 at 627-628. P. Drayton to S.F. Du Pont, May 11, 1863, ORN Ser. 1, Vol. 13 at 3-4.

Squadron commanders took the departure: D.G. Farragut to H. Eagle, Jul. 30, 1862, ORN Ser. 1, Vol. 19 at 101. D.G. Farragut to C. Hunter, Aug. 2, 1862, ibid. at 108. General Orders No. 6, Oct. 13, 1864, ORN Ser. 1, Vol. 10 at 561.

The ration provided by the Navy: E. Barnes and J.A. Barnes, eds., Naval Surgeon at 154 and *passim*.

By 1860 using ice to preserve: O.E. Anderson, Jr., Refrigeration in America at 3-96. Report of the Secretary of the Treasury, Transmitting a Report from the Register of the Treasury of the Commerce and Navigation of the United States for the Year Ending June 30, 1860 at 24-25.

Early in the war, the Federal Navy: O.E. Anderson, Jr., Refrigeration in America at 3-96. Report of the Secretary of the Treasury, Transmitting a Report from the Register of the Treasury of the Commerce and Navigation of the United States for the Year Ending June 30, 1860 at 24-25.

The arrangements for preserving: M. Woodhull to G. Welles, Oct. 4, 1861, ORN Ser. 1, Vol. 27 at 367-368. G. Welles to C.K. Stribling and enclosures, Jun. 5, 1863, ibid. at 502-505. W.B. Eaton to G. Welles, Aug. 5, 1864, ibid. at 606. W.B. Eaton to G. Welles, Sep. 25, 1864, ibid. at 622-623.

Water was a supply concern: M. Woodhull to G. Welles, Sep. 28, 1861, ORN Ser. 1, Vol. 27 at 366.

In addition to supplies of fresh beef: M. Woodhull to G. Welles, Feb. 19, 1862, ORN Ser. 1, Vol. 27 at 417. J.J. Almy to G. Welles, Aug. 23, 1864, ibid. at 612. J.E. Rockwell to G. Welles, Jan. 12, 1862, ibid. at 650. M. Woodhull to H. Paulding, Jan. 4, 1862, ibid. at 397.

With the majority of blockading vessels: G. Welles to W. Mervine, Jul. 8, 1861, ORN Ser. 1, Vol. 27 at 356. Endorsement of L.M. Goldsborough, May 1, 1862, ORN Ser. 21, Vol. 7 at 298-299. L.M. Goldsborough to J.B.M. Clitz, May 2, 1862, ibid. at 302. L.M. Goldsborough to G. Welles, May 5, 1862, ibid. at 322.

Civilian appointees in principal northern: D.G. Farragut to G. Welles, Mar. 6, 1862, ORN Ser. 1, Vol. 18 at 49-50. G. Welles to D.G. Farragut, Mar. 12, 1862, ibid. at 58. S.F. Du Pont to G. Welles, Mar. 27, 1862, J.D. Hayes, ed., Samuel Francis Du Pont: A Selection from His Civil War Letters, Vol. 1 at 391. G. Welles to S.F. Du Pont, Apr. 3, 1862, ORN Ser. 1, Vol. 12 at 707. S. F. Du Pont to J. Lenthall, Sep. 1, 1862, S.F. Du Pont to G. Welles, Sep. 26, 1862, in J.D. Hayes, ed., Samuel Francis Du Pont: A Selection from His Civil War Letters, Vol. 2 at 214 and 236. D.G. Farragut to H.A. Adams, May 7, 1864, ORN Ser. 1, Vol. 21 at 264. J.D. Hayes, ed., Samuel Francis Du Pont: A Selection from His Civil War Letters, Vol. 1 at 383 n. 11. A.A. Harwood to E.P. McCrea, Nov. 30, 1862, ORN Ser. 1, Vol. 5 at 175. D.G. Farragut to G. Welles, Jul. 15, 1864, ORN Ser. 1, Vol. 21 at 375.

Blockading tactics affected the demand: R.M. Browning, Jr., Success Is All That Was Expected at 108, 115-118 and 261-262. W.W. McKean to G. Welles, Oct. 25, 1861, ORN Ser. 1, Vol. 16 at 707. W.W. McKean to G. Welles, Nov. 1, 1861, ibid. at 752. E.g., J.J. Almy to S.F. Du Pont, Aug. 22, 1862, ORN Ser. 1, Vol. 13 at 252. J.A. Dahlgren to G. Welles, Jul. 29, 1863, ORN Ser. 1, Vol. 14 at 406. R.M. Browning, Jr., From Cape Charles to Cape Fear at 189-190 and 225. General Instructions of S.P. Lee, Dec. 16, 1863, ORN, Ser. 1, Vol. 9 at 356-357. S.P. Lee to O.S. Glisson, Jul. 9, 1864, ORN Ser. 1, Vol. 10 at 244.

The growth of the squadrons required: W. Reynolds to S.F. Du Pont, Feb. 16, 1863, ORN Ser. 1, Vol. 13 at 667-668.

The *Vermont* and her sister ship: Dictionary of American Naval Fighting Ships, New Hampshire I (Ship-of-the-Line), https://www.history.navy.mil/research/histories/ship-histories/danfs/n/new-hampshire-i.html. Ibid. Vermont I (SL), https://www.history.navy.mil/research/histories/ship-histories/danfs/v/vermont-i.html.

Blockading was wearing duty: J.S. Corbett, Principles of Maritime Strategy at 190-191.

The vessel's engineers determined: W. Chandler to G. Welles, June 23, 1861, ORN Ser. 1, Vol. 5 at 736. Report to W. Chandler, Jun. 15, 1861, ORN Ser. 1, Vol. 5 at 724.

The Federal Navy Department provided: S.F. Du Pont to G.V. Fox, Nov. 12, 1861, ORN Ser. 1, Vol. 12 at 341-342. S.F. Du Pont to C.S. Boggs, Feb. 24, 1862, ibid. at 561. L.M. Goldsborough to W. Smith, Nov. 23, 1861, ORN Ser. 1, Vol. 6 at 452. A. Murray to S.P. Lee, Mar. 4, 1863, ORN Ser. 1, Vol. 8 at 587. S.P. Lee to H.K. Davenport, Mar. 20, 1863, ibid. at 617. W.W. McKean to G. Welles, Apr. 19, 1862, ORN Ser. 1, Vol. 17 at 217. A. Read to D.G. Farragut, May 10, 1862, ORN Ser. 1, Vol. 18 at 848. S.P. Lee to G. Welles, Apr. 9, 1864, ORN Ser. 1, Vol. 9 at 591. D.G. Farragut to J. Smith Aug. 20, 1862, ORN Ser. 1, Vol. 19 at 163. D.G. Farragut to G. Welles, Aug. 20, 1862, ibid. at 164.

The resources at the blockading squadrons: Testimony of G.V. Fox, Mar. 19, 1862, in Report of the Joint Committee on the Conduct of the War (Monitor and Merrimack) at 415. S.F. Du Pont to S. Du Pont, May 1, 1862; S.F. Du Pont to J. Lenthall, Sep. 1, 1862; S.F. Du Pont to S. Du Pont, Aug. 24, 1862; S.F. Du Pont to S. Du Pont, Sep. 7, 1862; S.F. Du Pont to S. Du Pont, Sep. 12, 1862, in J.D. Hayes, ed., Samuel Francis Du Pont: A Selection From His Civil War Letters, Vol. 2 at 23, 198, 214, 219 and 226.

As the fleet grew and the need: W.W. McKean to G. Welles, Apr. 19, 1862, ORN Ser. 1, Vol. 17 at 217. T. Bailey to G. Welles, Aug. 5,

1864, ibid. at 760-761. S.F. Du Pont to C.S. Boggs, Feb. 24, 1862, ORN Ser. 1, Vol. 12 at 561. R.M. Browning Jr., Success Is All That Was Expected at 78. D.G. Farragut to I. Hanscom, May 19, 1864, ORN Ser. 1, Vol. 21 at 287.

Admiral Dahlgren reported in January: J.A. Dahlgren to G. Welles, Jan. 7, 1864, ORN Ser. 1, Vol. 15 at 225.

The British, mindful of the wearing: J.S. Corbett, Principles of Maritime Strategy at 190-191. G. Welles to D.G. Farragut, Dec. 12, 1862, ORN Ser. 1, Vol. 19 at 403. S.P. Lee to B.F. Garvin, Jul. 8, 1863, ORN Ser. 1, Vol. 9 at 118. J.A. Dahlgren to G. Welles, Jul. 24, 1863, ORN Ser. 1, Vol. 14 at 390. J.A. Dahlgren to G. Welles, Aug. 18, 1863, ibid. at 453. J.A. Dahlgren to G. Welles, Sep. 2, 1863, ibid. at 532. S.P. Lee to G. Welles, Mar. 25, 1864, ORN Ser. 1, Vol. 9 at 566-567. E.W. Sloan III, Benjamin Franklin Isherwood Naval Engineer at 189-190.

In March 1864 Admiral Lee wrote: S.P. Lee to G. Welles, Mar. 25, 1864, ORN Ser. 1, Vol. 9 at 566-567. E.W. Sloan III, Benjamin Franklin Isherwood Naval Engineer at 189-190.

The Confederates launched cruisers: G. Welles to S.F. Du Pont, Jan. 31, 1863, ORN Ser. 1, Vol. 13 at 571. G. Welles to G.J. Pendergrast, May 17, 1861, ORN Ser. 1, Vol. 5 at 636. G. Welles to G.J. Pendergrast, Aug. 29, 1861, ORN Ser. 1, Vol. 6 at 145-146. G. Welles to C. Wilkes, Sep. 8, 1862, ORN Ser. 1, Vol. 1 at 470-471. G. Welles to C. Wilkes, Dec. 15, 1862, ORN Ser. 1, Vol. 1 at 588.

In addition to imposing the blockade: E.g., T. Bailey to G. Welles (and attachments), Apr. 21, 1863, ORN Ser. 1, Vol. 17 at 390-394 (cutting out operations); E. English to J.L. Lardner (and attachments), Oct. 7, 1862, ibid. at 316-319 (destruction of salt works).

The composition of the Confederate Navy: Report of the Secretary of the Navy, Apr. 26, 1861, ORN Ser. 2, Vol. 2 at 51-52. Report of the Secretary of the Navy, Nov. 5, 1864, ORN Ser. 2, Vol. 2 at 743-747. The dates of two of Secretary Mallory's reports as presented in the Official Record appear to be in error. The Report dated April 30, 1864, ORN Ser. 2, Vol. 2 at 630 *et seq.*, makes reference to the sinking of the *Alabama* by the *Kearsarge*, which occurred in June 1864, and to the attack by Admiral Farragut on Mobile Bay and the capture of the *Tennessee*, which occurred in August 1864, in both cases after the stated April 30 date of the Report. The Report dated November 5, 1864, ORN Ser. 2, Vol. 2 at 743 *et seq.*, lists both the *Alabama* and the *Tennessee* as being in commission in Confederate service.

The North Atlantic Blockading Squadron: S.P. Lee to G. Welles, Jun. 17, 1864, ORN Ser. 1, Vol. 10 at 157-158.

The South Atlantic Blockading Squadron: Distribution of Vessels, Jun. 1, 1864, ORN Ser. 1, Vol. 15 at 465-466.

The East Gulf Blockading Squadron: Stations of Vessels, Jun. 1, 1864, ORN Ser. 1, Vol. 17 at 713-714. G. Welles to C. Wilkes, Sep. 8, 1862, ORN Ser. 1, Vol. 1 at 470.

The West Gulf Blockading Squadron: D.G. Farragut to G. Welles, Jun. 24, 1864, ORN, Ser. 1, Vol. 21 at 342-343. P.H. Silverstone, Civil War Navies at 111.

Chapter 6: Ramparts, Raiders and Rams

Initially, the Confederates responded: J.M. McPherson, Battle Cry of Freedom at 302-304 and 337. R.E. Lee to S. Cooper, Nov. 21, 1861, ORA Ser. 1, Vol. 6 at 327. R.E. Lee to S. Cooper, Jan. 8, 1862, ibid. at 367. R. Reed, Combined Operations in the Civil War at 48. W.H. Roberts, Now for the Contest at 35 and 60-61.

The principal assets that the Confederates: Circular of Instructions of R.S. Ridley, Dec. 26, 1862, ORN Ser. 1, Vol. 14 at 102-105.

At the major Confederate ports: C.E. Fonvielle, Jr., The Wilmington Campaign at 43-45. R. Gragg, Confederate Goliath at 18-21 and illustration. C.B. Comstock to R. Delafield, Jan. 23, 1865, ORA Ser. 1, Vol. 46, Pt. 2 at 215-216.

For a nation at war: H. Wheaton, Elements of International Law at 477-482. R. Phillimore, Commentaries upon International Law, Vol. 3 at 344.

The possession of letters of marque: Proclamation by J. Davis, Apr. 17, 1861, ORN Ser. 2, Vol. 3 at 96-97. An act recognizing the existence of war between the United States and the Confederate States, and concerning letters of marque, prizes and prize goods, May 6, 1861, ORN Ser. 2, Vol. 1 at 338.

While many seafaring nations authorized: J.T. Scharf, History of the Confederate States Navy at 58 and note 2. Lord J. Russell to C.F. Adams, Aug. 19, 1861, and Draft of Declaration, Message of the President of the United States to the Two Houses of Congress at the Commencement of the Second Session of the Thirty-Seventh Congress, Vol. 1 at 134

Early in the war, the Federal: Proclamation of A. Lincoln, Apr. 19, 1861, ORN Ser. 1, Vol. 4 at 156-157. J. Davis to A. Lincoln, Jul. 6, 1861, ORN Ser. 2, Vol. 3 at 103-104. W.M. Robinson, Jr., The Confederate Privateers, at 26 and 150-151. Lord J. Russell to Lords Commissioners of the Admiralty, Jun. 1, 1861, quoted in "The American Question", *New York Times*, Jun. 18, 1861. See also Report of R. Semmes, Nov. 9, 1861, ORN Ser. 1, Vol. 1 at 633. J.W. Simson, Naval Strategies of the Civil War at 187. See also P.H. Silverstone, Civil War Navies at 193-194. E.S Maclay, A History of American Privateers at 506. List of privateers known to have been commissioned by the Confederate States, together with a list of their prizes, ORN Ser. 1, Vol. 1 at 818-819.

During 1862 and 1863: J. Niven, Gideon Welles at 448-449 and 451. An Act concerning Letters of Marque, Prizes and Prize Goods, Mar. 3, 1863, U.S. Statutes at Large, Vol. 12 at 758. C. Sumner to A. Lincoln, Mar. 18, 1863, Library of Congress, Abraham Lincoln Papers, http://memory.loc.gov/ammem/alhtml/malhome.html.

G. Welles to W.H. Seward, Mar. 31, 1863, in G. Welles, Diary of Gideon Welles Vol. 1 at 253.

Unlike a privateer, a cruiser: R.W. Tucker, The Law of War and Neutrality at Sea at 106-107. R. Luraghi, A History of the Confederate Navy at 422 in note 71. H. Wheaton, Elements of International Law at 506-508. R. Phillimore, Commentaries upon International Law, Vol. 3 at 379-380.

The Confederates sought cruisers: C.G. Hearn, Gray Raiders of the Sea at 10-14. C.H. Poor to W. Mervine, Jun. 30, 1861, ORN Ser. 1, Vol. 1 at 34. S.R. Mallory to J.D. Bulloch, May 9, 1861, ORN Ser. 2, Vol. 2 at 64-65. H. Wheaton, Elements of International Law at 532-533. R. Phillimore, Commentaries Upon International Law, Vol. 3 at 180-183. E.g., C.F. Adams to Lord J. Russell, Feb. 18, 1862, Message of the President of the United States to the Two Houses of Congress at the Commencement of the Third Session of the Thirty-Seventh Congress, Vol. 1 at 39. L.M. Case and W.F. Spencer, The United States and France at 436-54. J. Bigelow, France and the Confederate Navy 1862-1868 at 1-55.

Purchase was the preferred Confederate: Case of the United States laid before the Tribunal of Arbitration at Geneva, U.S. Department of State, Papers Relating to the Treaty of Washington, Vol. 1 at 135-136. Case presented on the part of the Government of Her Britannic Majesty to the Tribunal U.S. Department of State, Papers Relating to the Treaty of Washington, Vol. 1 at 315.

Unable to find any suitable vessels: J.D. Bulloch, The Secret Service of the Confederate States in Europe at 39, 41 and 43-44. H. Wheaton, Elements of International Law at 523-533. R. Phillimore, Commentaries Upon International Law Vol. 3 at 180-183

One of the sharpest continuing points: H. Wheaton, Elements of International Law at 532-533. R. Phillimore, Commentaries Upon International Law, Vol. 3 at 180-183.

The actions of private citizens: Foreign Enlistment Act, 59 Geo. III, Ch. 69 (1819), quoted in "Declaration of the British Government", May 14, 1861, *New York Times*, May 29, 1861. J.D. Bulloch, The Secret Service of the Confederate States in Europe at 46-48, 114-115, 168-171 and 181. C.C. Beaman, Jr. The National and Private Alabama Claims and their Final and Amicable Settlement at 63 and 66-67.

As the depredations of the British-built: Lord J. Russell to Lord Lyons, Mar. 28, 1863, quoted in Lord Newton, Lord Lyons: A Record of British Diplomacy, Vol. 1. at 99-100. Lord Lyons to Lord J. Russell, Apr. 24, 1863, quoted in ibid. at 102. J.D. Bulloch, The Secret Service of the Confederate States in Europe at 244. C.C. Beaman, Jr., The National and Private Alabama Claims and their Final and Amicable Settlement at 150-156.

During the course of the war: P.H. Silverstone, Civil War Navies. H.C. Blake to G. Welles, Jan. 21, 1863, ORN Ser. 1, Vol. 2 at 18-20. J.A. Winslow to G. Welles, Jun. 19, 1864, ORN Ser. 1, Vol. 3 at 59. N. Collins to G. Welles, Oct. 31, 1864, ibid. at 254-55. C.G. Hearn, Gray Raiders of the Sea at 148-149 and 261-301. R.E. Lee to J.A. Seddon, Sep. 24, 1864, ORN Ser. 1 Vol. 10 at 747. R.M. Browning, Jr., From Cape Charles to Cape Fear at 239.

The ability to strike the enemy directly: G. Welles to C. Wilkes, Dec. 15, 1862, ORN Ser. 1, Vol. 1 at 588.

Indeed, commissioning cruisers proved: R.E. Lee to J.A. Seddon, Sep. 24, 1864, ORN Ser. 1, Vol. 10 at 747. R.M. Browning, Jr., From Cape Charles to Cape Fear at 239.

People knowledgeable about naval: S.R. Mallory to J.H. North, May 17, 1861, ORN Ser. 2, Vol. 2 at 70-72. W.H. Roberts, Now for the Contest at 20. J.H. North, Aug. 16, 1861, ORN Ser. 2, Vol. 2 at 87.

With a monopoly—or a local monopoly: E.M. Stanton to the Governors of New York, Massachusetts and Maine, Mar. 9,

1862, ORN Ser. 1, Vol. 7 at 80. J.A. Dahlgren to J. Hooker, Mar. 9, 1862, ibid. at 75.

All but one of the Confederate-built ironclads: P.H. Silverstone, Civil War Navies at 152-157. J.T. Scharf, History of the Confederate States Navy at 152.

The Confederates' efforts to build: W.H. Roberts, Now for the Contest at 80-81.

While the Confederate ironclads enjoyed: W.C. Davis, Duel Between the First Ironclads at 116-137. P.H. Silverstone, Civil War Navies at 152-154. H.S. Stellwagen to S.F. Du Pont, Jan. 31, 1863, ORN Ser. 1, Vol. 13 at 579-580. S.F. Du Pont to G. Welles, Feb 1, 1863, ORN Ser. 1, Vol. 13 at 577-578. J.R. Soley, The Blockade and the Cruisers at 116-118. D.G. Farragut to G. Welles, Aug. 5, 1864, ORN Ser. 1, Vol. 21 at 405. D.G. Farragut to G. Welles, Aug. 12, 1864, ORN Ser. 1, Vol. 21 at 418 and 420. J. Alden to D.G. Farragut, Aug. 6, 1864, ORN Ser. 1, Vol. 21 at 445-446.

As the war entered its second year: J.D. Bulloch to S.R. Mallory, Apr. 11, 1862, ORN Ser. 2, Vol. 2 at 184. J.D. Bulloch to S.R. Mallory, Jul. 21, 1862, ibid. at 223-26. S.C. Tucker, A Short History of the Civil War at Sea at 103. J. and G. Thomson to J.D. North, May 7, 1862, ORN Ser. 2, Vol. 2 at 191-92. S.R. Mallory to J.D. Bulloch, May 6, 1863, ibid. at 416. S.R. Mallory to J.D. Bulloch, May 26, 1863 ibid. at 428-429. F.J. Merli, Great Britain and the Confederate Navy at 154. L.M. Case and W.F. Spencer, The United States and France at 436-54 and 466-80. J. Bigelow, France and the Confederate Navy 1862-1868 at 1-55. J.W. Simson, Naval Strategies of the Civil War at 166. R. Luraghi, A History of the Confederate Navy at 343-344. J. Bigelow, France and the Confederate Navy 1862-1868 at 56-89. W.F. Spencer, The Confederate Navy in Europe at 196-207.

The Confederates sought novel: J.S. Barnes, Submarine Warfare, Offensive and Defensive at 17-29 and 57-60. S.C. Rowan to G.

Welles and enclosures, Jul. 7, 1861, ORN Ser. 1, Vol. 4 at 566-568. M.P. Kochan and J.C. Wideman, Torpedoes at 37-74. J.A. Dahlgren to G. Welles, Mar. 21, 1865, ORN Ser. 1, Vol. 16 at 296.

Local defense forces designed: J.S. Barnes, Submarine Warfare, Offensive and Defensive at 63-78. S.F. Du Pont to G. Welles, Apr. 15, 1863, ORN Ser. I, Vol. 14 at 6. P. Drayton to S.F. Du Pont, Apr. 8, 1863, ibid. at 9. J. Rodgers to S.F. Du Pont, Apr. 8, 1863, ibid. at 12. Compare J. Johnson, The Defense of Charleston Harbor at 32 (torpedoes not deployed before July 1863) with Examination of M.M. Gray, Apr. 11, 1865, ORN Ser. 1, Vol. 16 at 412 (torpedoes were deployed starting in February 1863). G. Welles to J.A. Dahlgren, Nov. 2, 1863, ORN Ser. 1, Vol. 15 at 96. M.P. Kochan and J.C. Wideman, Torpedoes at 96. G.J. Rains to J.A. Seddon, Aug. 15, 1864, ORN Ser. 1, Vol. 21 at 567. J.B. Marchand to D.G. Farragut, Feb. 18, 1864, ORN Ser. 1, Vol. 21 at 105-106. D.H. Maury, Sep. 26, 1864, ORN Ser. 1, Vol. 21 at 570. D.H. Maury, Sep. 26, 1864, ORN Ser. 1, Vol. 21 at 570. Statement of J.R. Patrick, Jun. 27, 1864, ORN Ser. 1, Vol. 9 at 770.

The Federals tried different strategies: W.W.W. Wood to G. Welles, Apr. 16, 1863, ORN Ser. 1, Vol. 14 at 138. Sketch of "Devil", Apr. 7, 1863, ORN Ser. 1, Vol. 14 at 93. D.G. Farragut to T.A. Jenkins, Aug. 3, 1864, ORN Ser. 1, Vol. 21 at 403. General Order of D.D. Porter, Oct. 13, 1864, ORN Ser. 1, Vol. 10 at 562. D.D. Porter to E.T. Nichols, Oct. 28, 1864, ORN Ser. 1, Vol. 11 at 4. D.D. Porter to W.A. Parker, Dec. 2, 1864, ibid. at 120. W.H. Macomb to D.D. Porter, Dec. 11, 1864, ORN Ser. 1, Vol. 11 at 161. H.L. Wait, "The Blockade of the Confederacy", The Century, Oct. 1898, Vol. 56, Issue 6 at 924. Report of R. Chandler and enclosures, May 24, 1863, ORN Ser. 1, Vol. 12 at 98-99.

The Confederates defended their harbors: J.S. Barnes, Submarine Warfare, Offensive and Defensive at 64-78. P.G.T. Beauregard to R.S. Ripley, Mar. 18, 1863, ORA Ser. 1, Vol. 14 at 835. Addendum to Statement of W. Smith, Jun. 2, 1865, ORN Ser. 1, vol. 16 at 411 (obstructions). Examination of M.M. Gray, Apr. 11, 1865, ibid. at 412 (torpedoes).

The United States had employed: S.C. Tucker, The Jeffersonian Gunboat Navy. S.C. Tucker, Blue & Gray Navies at 44. M.F. Maury to W.B. Preston, Oct. 22, 1861, ORN Ser. 2, Vol. 2 at 103. J.B. Hull to G. Welles, Aug. 30, 1861, ORN Ser. 1, Vol. 6 at 148. M.F. Maury to W.B. Preston, Oct. 22, 1861, ORN Ser. 2, Vol. 2 at 99-104. A bill to authorize the President to cause to be constructed a certain number of gunboats, Dec. 23, 1861, ibid. at 117. An act making appropriations for one hundred gunboats, Dec. 23, 1861, ibid. at 117. M.F. Maury to S.R. Mallory, Jan. 25, 1862, ORN Ser. 2, Vol. 2 at 138.

The torpedo boat was a more: M.F. Perry, Infernal Machines at 64.

In anticipation of a Federal ironclad: W.H. Parker, Recollections of a Naval Officer 1841-1865 at 332-340.

Nighttime attacks with steam-powered: J. Carlin to P.G.T. Beauregard, Aug. 22, 1863, ORN Ser. 1, Vol. 14 at 498-499. P.G.T. Beauregard to J. Carlin, Aug. 23, 1863, ibid. at 500. J.H. Tomb to J.R. Tucker, Oct. 6, 1863, ORN Ser. 1, Vol. 15 at 20-21. J.A. Dahlgren to G. Welles, Nov. 19, 1863, ibid. at 16-17. J.A. Dahlgren to G. Welles and enclosures, No. 30, 1863, ibid. at 17-18.

The *New Ironsides*, the largest: P.H. Silverstone, Civil War Navies at 11 and 165. J.A. Dahlgren to G.V. Fox and enclosure, ORN Ser. 1, Vol. 15 at 14.

The *David* attempted other attacks: R.O. Patterson to S.C. Rowan, Mar. 6, 1864, ORN Ser. 1, Vol. 15 at 356-357. F.D. Lee to T. Jordan, Mar. 8, 1864, ibid. at 358. Extract of notebook of J.H. Tomb, no date, ibid. at 358-359. J. De Camp to S.C. Rowan, Apr. 19, 1864, ORN ibid. at 405. Extract of notebook of J.H. Tomb, no date, ibid. at 359. S.P. Lee to G. Welles and enclosures, Apr.

15, 1864, ORN Ser. 1, Vol. 9 at 592-594. P.H. Silverstone, Civil War Navies at 165-167. S.R. Mallory to J.D. Bulloch and enclosure, Apr. 16, 1864, ORN Ser. 2, Vol. 2 at 627-28. S.R. Mallory to J.D. Bulloch, Jul. 18, 1864, ibid. at 688-689. J.D. Bulloch to S.R. Mallory, Jul. 8, 1864, ibid. at 683. S.R. Mallory to J.D. Bulloch and enclosure, Apr. 16, 1864, ibid. at 627-28. J.D. Bulloch to S.R. Mallory, Jul. 8, 1864, ibid. at 683. J.D. Bulloch to S.R. Mallory, Oct. 20, 1864, ibid. at 735. J.D. Bulloch to S.R. Mallory, Jan. 26, 1865, ibid. at 790-791.

Most of the torpedo boat activity: R.B. Hitchcock to T.A. Jenkins, Feb. 25, 1863, ORN Ser. 1, Vol. 19 at 631. C.H. Greene to H.H. Bell, Nov. 19, 1863, ORN Ser. 1, Vol. 20 at 690-691. H.H. Bell to C.H. Greene, Nov. 25, 1863, ibid. at 697. Special Orders No. 13, Jan. 28, 1865, ORN Ser. 1, Vol. 11 at 708. W.B. Cushing to D.D. Porter, Oct. 30, 1864, ORN Ser. 1, Vol. 10 at 611-612. T.S. Gay to G. Welles, Mar. 7, 1865, ORN Ser. 1, Vol. 10 at 613-614.

In addition to surface craft and: P.H. Silverstone, Civil War Navies at 25 and 167. Illustration of Confederate States Submarine Torpedo Boat H.L. Hunley, ORN Ser. 1, Vol. 15 after 338. M.F. Perry, Infernal Machines at 99. F.J. Higginson to J.A. Dahlgren, Feb. 18, 1864, ORN Ser. 1, Vol. 15 at 328. J.A. Dahlgren to G. Welles, Feb. 19, 1864, ibid. at 329-330. M.M. Gray to D.H. Maury, Apr. 19, 1864, ibid. at 337-338.

The Federals countered the threat: Examination of M.M. Gray, Apr. 11, 1985, ORN Ser. 1, Vol. 16 at 413. J.A. Dahlgren to G. Welles, Jan. 13, 1864, ORN Ser. 1, Vol. 15 at 238-239. S.P. Lee to W.H. Macomb and enclosure, Jul. 20, 1864, ORN Ser. 1, Vol. 10 at 293-294. D.D. Porter to W.A. Parker, Dec. 2, 1864, ORN Ser, 1, Vol. 11 at 120. Special order of D.D. Porter, Jan. 28, 1865, ibid. at 708. Special order to D.D. Porter, Feb. 25, 1865, ORN Ser. 1, Vol. 12 at 51-53.

Chapter 7: Jealously Guarded Prerogatives
The conditions for achieving victory: W.L. Yancey and A.D. Mann to R. Toombs, May 21, 1861, ORN Ser. 2, Vol. 3 at 215-216. W.L. Yancey, P.A. Rost and A.D. Mann to Lord J. Russell, Aug. 14, 1861, ORN Ser. 2, Vol. 3 at 246. J. Slidell to E. Thouvenel, Jul. 21, 1862, ORN Ser. 2, Vol. 3 at 476. J.M. Mason to Lord J. Russell, Aug. 1, 1862, ORN Ser. 2, Vol. 3 at 501-503. R.M.T. Hunter to J.M. Mason, Sep. 23, 1861, ORN Ser. 2, Vol. 3 at 260. R.M.T. Hunter to J. Slidell, Sep. 23, 1861, ORN Ser. 2. Vol. 3 at 268. J.P. Benjamin to J.M. Mason, Apr. 12, 1862, ORN Ser. 2, Vol. 3 at 386. J.P. Benjamin to J. Slidell, Apr. 12, 1862, ORN Ser. 2, Vol. 3 at 389-390.
International law stated that premature recognition: W.G.G. Harcourt, Letters by Historicus on Some Questions of International Law at 3-37.
On November 8, 1861, Captain Charles Wilkes: G.H. Warren, Fountain of Discontent at 13 and 16-25. D.P. Crook, The North, the South, and the Powers at 103-106. H. Wheaton, Elements of International Law at 604-605 and 630-36. R. Phillimore, Commentaries Upon International Law, Vol. 3 at 239 and 284. H. Jones, Blue & Gray Diplomacy at 97. D.B. Mahin, One War at a Time at ix and 258-259.
Although the Federal government backed down: R. W. Winks, The Civil War Years at 105-110.
France invaded Mexico in late 1861: A.J. Hanna and K.A. Hanna, Napoleon III and Mexico at 40 and 43-46.
Britain's failure to prevent the sailing: Lord J. Russell to Lord Lyons, Mar. 28, 1863, quoted in Lord Newton, Lord Lyons: A Record of British Diplomacy, Vol. 1 at 99-100. Lord Lyons to Lord J. Russell, Apr. 24, 1863, quoted in ibid. at 102. J.D. Bulloch, The Secret Service of the Confederate States in Europe at 244. C.C. Beaman, Jr., The National and Private Alabama Claims and

their Final and Amicable Settlement at 150-156. J.D. Bulloch, The Secret Service of the Confederate States in Europe at 315. Memorandum accompanying J. Slidell to J.P. Benjamin, Oct. 28, 1862, ORN Ser. 2, Vol. 3 at 576-577. See also J. Slidell to J.P. Benjamin, Mar. 4, 1863, ORN Ser. 2, Vol. 3 at 706; J. Slidell to J.P. Benjamin, Apr. 11, 1863, ORN Ser. 2, Vol. 3 at 738. L.M. Case and W.F. Spencer, The United States and France at 436-54 and 466-480. J. Bigelow, France and the Confederate Navy 1862-1868 at 1-89. R. Luraghi, A History of the Confederate Navy at 343-344.

Britain actively considered, but ultimately: H. Jones, Blue & Gray Diplomacy at 254 and 270-270. F.L. Owsley, King Cotton Diplomacy at 546.

Although President Lincoln's preliminary Emancipation: H. Hotze to J.P. Benjamin, Oct. 24, 1862, ORN Ser. 2, Vol. 3 at 566-567. W.G.G. Harcourt, Letters by Historicus on Some Questions of International Law at 49. H. Martin, Britain in the 19th Century at 102. J.M. Mason to J.P. Benjamin, Nov. 7, 1862, ORN Ser. 1, Vol. 3 at 600. "Voices of British Workingmen" From the *London Daily News*, Jan. 2, 1863, in Littell's Living Age, Vol. 76 (1863) at 329-330. C.F. Adams to W.H. Seward, Jan. 16, 1863, and enclosures, Messages of the President of the United States and Accompanying Documents, 38th Cong. 1st Sess., Vol. 1 at 59-63. B. Moran to W.H. Seward, Jan. 17, 1863, and enclosure, ibid. at 63-66. C.F. Adams to W.H. Seward, Jan. 22, 1863, and enclosures, ibid. at 92-96. C.F. Adams to W.H. Seward, Feb. 5, 1863, ibid. at 115. C.F. Adams to W.H. Seward, Feb. 12, 1863, and enclosures, ibid. at 128-131.

In July 1864 Lord Russell: Lord J. Russell to Lord Lyons, Jul. 23, 1864, quoted in Lord Newton, Lord Lyons: A Record of British Diplomacy, Vol. 1 at 132-133.

International law viewed the oceans: J.T. Abdy, ed., Kent's Commentaries on International Law at 143-144, 97-98, 246-249 and 399-400. H. Wheaton, Elements of International Law at 193, 426-428, 509 and 520-521.

Neutrals retained a conditional right: R. Phillimore, Commentaries Upon International Law, Vol. 3 at 250, 252 and 285. H. Wheaton, Elements of International Law at 607-609 and 638-644. J.T. Abdy, ed., Kent's Commentaries on International Law at 303 and 330.

A key question was what constituted: R. Phillimore, Commentaries Upon International Law, Vol. 3 at 257 and 277. H. Wheaton, Elements of International Law at 610-637.

A belligerent also could impose a blockade: R. Phillimore, Commentaries Upon International Law, Vol. 3 at 291-292, 299 and 305-396. J.T. Abdy, ed., Kent's Commentaries on International Law at 340-344 and 350-351. H. Wheaton, Elements of International Law at 668-670.

The existence of a lawful blockade: R. Phillimore, Commentaries Upon International Law, Vol. 3 at 253-256 and 305-396. H. Wheaton, Elements of International Law at 638-644. J.T. Abdy, ed., Kent's Commentaries on International Law at 343-344 and 350-351.

Early in 1862 Britain declared that: F.L. Owsley, King Cotton Diplomacy at 222-23. J.P. Benjamin to J.M. Mason, Apr. 8, 1862, ORN Ser. 2, Vol. 3 at 380.

As the Civil War began: An Act further to provide for the Collection of Duties on Imports, Sec. 3, U.S. Statutes at Large, Vol. 12 at 256-257. Lord Lyons to Lord J. Russell, Apr. 15, 1861, quoted in Lord Newton, Lord Lyons: A Record of British Diplomacy, Vol. 1. at 36. Report of the Secretary of the Navy, Jul. 4, 1861, Congressional Globe App., 37th Cong. 1st Sess. at 8. G. Welles to

NOTES

A. Lincoln, Aug. 5, 1861, ORN Ser. 1, Vol. 6 at 53-56. G. Welles to S.H. Stringham, Aug. 19, 1861, ORN Ser. 1, Vol. 6 at 93.

From that point, Secretary Welles: Annals of British Legislation, New Ser. Vol. 1 (1865) at 231-232. H.K. Beale, Diary of Gideon Welles, Vol. 1 at 82. Instructions from G. Welles, Aug. 18, 1862, ORN Ser. 1, Vol. 1 at 417-418; ORN Ser. 1, Vol. 7 at 656-657.

The British also sought an instruction: Annals of British Legislation, New Ser. Vol. 1 (1865) at 232-233. W.H. Seward to G. Welles, Oct. 31, 1882, Message of the President of the United States, and Accompanying Documents, to the Two Houses of Congress at the Commencement of the First Session of the Thirty-Eighth Congress at 456. An Act for the better government of the Navy of the United States, Apr. 23, 1800, United States Statutes at Large, Vol. 2 at 46. An Act for the better Government of the Navy of the United States, Jul. 17, 1862, United States Statutes at Large Vol. 12 at 607. F.T. Pratt, ed., Notes on the Principles and Practice of Prize Courts, by the Late Judge Story (1854) at 14-19. F.H. Upton, The Law of Nations Affecting Commerce During War with a review of the Jurisdiction, Practice and Proceedings of Prize Courts (1861) at 248-252. H. Wheaton, Elements of International Law at 630-636. R. Phillimore, Commentaries upon International Law, Vol. 3 at 284-285. The British Steam-ship "Peterhoff," a Report on her Seizure by the United States Cruiser "Vanderbilt," and the Subsequent Proceedings in the United States Prize Court, in New York (1863) at 16-17.

Not long after, a Federal cruiser: C. Wilkes to G. Welles, Feb. 25, 1863, ORN Ser. 1, Vol. 2 at 97-98. C.H. Baldwin to G. Welles, Feb. 25, 1863, ibid. at 98. H.K. Beale, ed., Diary of Gideon Welles, Vol. 1 at 270. Hansard's Parliamentary Debates, Vol. 152 (1863) at 765. The British Steam-ship "Peterhoff," a Report on her Seizure by the United States Cruiser "Vanderbilt," and the Subsequent

Proceedings in the United States Prize Court, in New York (1863) at 8-34. S.L. Bernath, Squall Across the Atlantic at 72-73.

The British government also sought: H.K. Beale, ed., Diary of Gideon Welles, Vol. 1 at 398. A. Lincoln to G. Welles, July 25, 1863, ORN Ser. 1, Vol. 2 at 411. J.B. Hattendorf, "The Royal Navy During the War of the French Revolution and the Napoleonic War" in D. King, A Sea of Words at 5.

When Secretary Welles received the British proposed: H.K. Beale, ed., Diary of Gideon Welles, Vol. 1 at 450-451. Memorandum, no date, ibid. at 454-466. G. Welles to A. Lincoln, Sep. 30, 1863, ibid. at 452-453.

Secretary Welles was less successful: G. Welles to S.P. Lee, Jan. 11, 1864, ORN Ser. 1, Vol. 9 at 405. G. Welles to D.G. Farragut, May 9, 1864, ORN Ser. 1, Vol. 10 at 61-62. G. Welles to S.P. Lee, ORN Ser. 1, Vol. 10 at 61. G. Welles to C.K. Stribling, Sep. 23, 1864, ORN Ser. 1, Vol. 17 at 759. S.L. Bernath, Squall Across the Atlantic at 142-149.

Toward the end of the war: "Disloyal Citizens and blockade-runners —Important order", Mar. 4, 1865, ORA Ser. 1, Vol. 47, Pt. 3 at 53-54.

The northern states were themselves: P. Leigh, Trading with the Enemy. "Circular to collectors, surveyors, and other officers of the customs", May 2, 1861, Report of the Joint Committee on the Conduct of the War (Trade in Military Districts), Part 3 at 561-562. An Act further to provide for the Collection of Duties on Imports, and for other Purposes, Jul. 13, 1861, Statutes at Large, Vol. 12 at 255-258.

These restrictions acquired a substantial: W.H. Seward to W. Stuart, Oct. 3, 1862, Message of the President of the United States to the Two Houses of Congress at the Commencement of the Third Session of the Thirty-Seventh Congress (1862), Vol. 1 at 302. "An Act supplementary to An Act to provide for the Collection of Duties on Imports, and for other purposes", May 20, 1862, Statutes at Large, Vol. 12 at 404-405. H. Barney to S.P. Chase and enclosures, Aug. 9, 1862, Message of the President of the United States to the Two Houses of Congress at the Commencement of the Third session of the Thirty-Seventh Congress (1862), Vol. 1 at 275-282. Percentage calculated from Report of the Secretary of the Treasury, Transmitting a Report from the Register of the Treasury of the Commerce and Navigation of the United States, for the Year Ended June 30, 1860 at 50-51, 316-317 and 350-351. L.H. Johnson, "Commerce Between Northeastern Ports and the Confederacy, 1861-1865", Journal of American History, Vol. 54, No. 1 (1967) at 32.

Federal legislation authorized the secretary: An Act supplementary to An Act to provide for the Collection of Duties on Imports, and for other purposes, May 20, 1862, Statutes at Large, Vol. 12 at 404-405. H. Barney to S.P. Chase and enclosures, Aug. 9, 1862, Message of the President of the United States to the Two Houses of Congress at the Commencement of the Third Session of the Thirty-Seventh Congress (1862), Vol. 1 at 275-282. Percentage calculated from Report of the Secretary of the Treasury, Transmitting a Report from the Register of the Treasury of the Commerce and Navigation of the United States, for the Year Ended June 30, 1860 at 50-51, 316-317 and 350-351.

The British government protested: W. Stuart to W.H. Seward, Aug. 1, 1862, Message of the President of the United States to the Two Houses of Congress at the Commencement of the Third Session of the Thirty-Seventh Congress (1862), Vol. 1 at 273-274. W. Stuart to W.H. Seward, Sep. 25, 1862, ibid. at 293-295. Lord J. Russell to W. Stuart, Sep. 22, 1862, ibid. at 304-306. Russell to W. Stuart, Dec. 17, 1862, Annals of British Legislation New Ser. Vol. 1 (1865) at 254-261.

The matter remained unresolved: H. Barney to S.P. Chase and enclosures, June 23, 1863, Message of the President of the United States, and Accompanying Documents, to the Two Houses of Congress, at the Commencement of the First Session of the Thirty-Eighth Congress, Vol. 1 (1864), at 674-677. H. Barney to S.P. Chase, May 27, 1863, ibid. at 677. W.H. Seward to Lyons, Jan. 7, 1864, Papers relating to foreign affairs, accompanying the annual message of the president to the second session thirty-eighth congress, Vol. 2 (1865) at 471-473. C.P. Clinch to W.P. Fessenden, Aug. 3, 1864, Papers relating to foreign affairs, accompanying the annual message of the president to the second session thirty-eighth congress, Vol. 2 (1865) at 683.

The Federal customs officials': R.W. Rawson, Report on the Bahamas for the Year 1864 at 50.

Both the British national government: S. Whiting to G. Welles, Dec. 16, 1861, ORN Ser. 1, Vol. 1 at 246. W.G. Temple to G. Welles and enclosures, Dec. 19, 1861, ibid. at 252-255. J. DeCamp to G. Welles, Dec. 22, 1861, ibid. at 257-258.

In January 1862 the British Foreign Office: Letter from Lord J. Russell to the Lords Commissioners of the Admiralty, Jan. 31, 1862, ORN Ser. 1, Vol. 1 at 325-327. C. Wilkes to J.L. Lardner, Jun. 20, 1863, ORN Ser. 1, Vol. 2 at 359.

Other nations did not impose similar: C. Wilkes to J.L. Lardner, Jun. 20, 1863, ORN Ser. 1, Vol. 2 at 359.

After the Civil War, the United States: A. Cook, The Alabama Claims. Opinions of Count Sclopis and Mr. Adams, no date, U.S. Department of State, Papers Relating to the Treaty of Washington, Vol. 4 at 74-75 and 148-150.

In addition to attempting to influence: J.F. Ross to W. W. McKean, Dec. 13, 1861, ORN Ser. 1, Vol. 17 at 10. W.W. McKean to G. Welles, Dec. 31, 1861, ibid. at 29. H.L. Wait, "The Blockade of the Confederacy", The Century, Oct. 1898, Vol. 56, Issue 6 at 921. Firing at the steamer *Herald*, Jul 25, 1862, ORN Ser. 1, Vol. I at 403-413. Treatment of crew of the schooner *Revere*, Sep. 10, 1861, ORN Ser. 1, Vol. 6 at 193; ORN Ser. 1, Vol. 12 at 407-410. Destruction of the British steamer *Night Hawk*, Sep. 29, 1864, ORN Ser. 1, Vol. 10 at 492-501. Capture of the British schooner *Telegraph*, Nov. 30, 1861, ORN Ser. 1, Vol. 16 at 797-800. Enforcement of the blockade with respect to the British steamer *Labuan* and the French steamer *Le Tage*, Feb. 1 and Mar. 17, 1862, ORN Ser. 1, Vol. 17 at 99-115; ORN Ser. 1, Vol. 18 at 77-86. Boarding of the British bark *Southport*, Apr. 7, 1862, ORN Ser. 1, Vol. 17 at 210-211. Capture of the British steamer *Adela*, Jul. 7, 1862, ibid. at 273-286. Firing upon the British schooner *Dream*, Apr. 28, 1863, ibid. at 426-428. Seizure of the British steamer *Sir William Peel*, Sep. 11, 1863, ORN Ser. 1, Vol. 20 at 465-467, 503-504 and 567-579. Seizure of vessels off Matamoras, 1863, ORN Ser. 1, Vol. 21 at 107-111.

Individual commanders of the naval vessels: C. Wilkes to G. Welles, Oct. 11, 1862, ORN Ser. 1, Vol. 1 at 500. T. Cochran, Blockade Runners of the Confederacy at 258. Journal of J.N. Maffet, May 4, 1862 *et seq.*, ORN Ser. 1, Vol. 1 at 764-765.

James Magee, the acting British consul: J.P. Benjamin to J.M. Mason, Jun 11, 1863, ORN Ser. 2. Vol. 3 at 798. R.A. Courtemanche, No Need of Glory at 174. Lord Lyons to Lord J. Russell, Nov. 17, 1863, quoted in Lord Newton, Lord Lyons: A Record of British Diplomacy, Vol. 1 at 122. Instructions from A. Milne, Feb. 16, 1863, ORN, Ser. 1, Vol. 19 at 616.

In a separate incident, the *Petrel*: S.F. Du Pont to G. Welles, Feb. 9, 1863, ORN Ser. 1, Vol. 13 at 601-602. T. Turner to S.F. Du Pont. Feb 8, 1863, ORN Ser. 1, Vol. 13 at 602-603. G. Welles to W.H. Seward, Feb 13, 1863, ORN Ser. 1, Vol. 13 at 662. J.H. Upshur to S.F. Du Pont, Feb. 23, 1863, ORN Ser. 1, Vol. 13 at 674. W.H. Seward to G. Welles, Feb. 13, 1863, ORN Ser. 1, Vol. 13 at 663. R.A. Courtemanche, No Need of Glory at 112-117.

Excess zeal in command was not: G. Welles to C. Wilkes, Dec. 15, 1862, ORN Ser. 1, Vol. 1 at 588. H.K. Beale, ed., Diary of Gideon Welles, Vol. 1 at 304-305, 309 and 322. G. Welles to J.L. Lardner, Jun. 1, 1863, ORN Ser. 1, Vol. 2 at 250-251. G. Welles to C. Wilkes, Jun. 1, 1863, ibid. at 253.

The profit opportunity created by: S.D. Trenchard to G. Welles, Jun. 1, 1863, ORN Ser. 1, Vol. 2 at 235-236.

The steamer was the *Margaret and Jessie*: L. Heyliger to J.P. Benjamin and enclosure, Jun. 6, 1863, ORN Ser. 1, Vol. 2 at 236-242. J. Fraser & Co. to J.P. Benjamin and enclosures, ibid. at 242-248. J.M. Mason to Lord J. Russell, Jul. 10, 1863, ibid. at 248. Lord J. Russell to J.M. Mason, Jul. 10, 1863, ibid. at 249.

A Federal Navy court of inquiry: Opinion of court of enquiry convened April 7, 1864, ORN Ser. 1, Vol. 2 at 249-250. J.H. Burnley to W.H. Seward, Sep. 10, 1864, Papers Relating to Foreign Affairs, Accompanying the Annual Message of the President to the Second Session Thirty-Eighth Congress, Vol. 2 (1865) 704-705. W.H. Seward to J.H. Burnley, Sep. 16, 1864, Papers Relating to Foreign Affairs, Accompanying the Annual Message of the President to the Second Session Thirty-Eighth Congress, Vol. 2 (1865) at 708-709.

The neutral powers insisted upon: G. Welles to C.K. Stribling, Sep. 23, 1864, ORN Ser. 1, Vol. 17 at 759.

Even though Confederate cotton shipped: H. Wheaton, Elements of International Law at 520-521, 551-572, 607-609 and 668-671. R. Phillimore, Commentaries upon International Law, Vol. 3 at 317.

In the fall of 1862 Federal officials: Letter dated Nov. 24, 1862, ORN Ser. 1, Vol. 19 at 419. W.H. Seward to G. Welles and endorsement,

Dec. 18, 1862, ORN Ser. 1, Vol. 19 at 418. C.F. Adams to W.H. Seward, May 1, 1863, ORN Ser. 1, Vol. 20 at 201. G. Welles to W.H. Seward, May 22, 1863, ORN Ser. 1, Vol. 17 at 446. G. Welles to W.H. Seward, Jun. 9, 1863, ORN Ser. 1, Vol. 20 at 291. D.G. Farragut to G. Welles, Mar. 27, 1862, ORN Ser. 1, Vol. 18 at 77. T. Bailey to G. Welles, Apr. 2, 1863, ORN Ser. 1, Vol. 17 at 403. G. Welles to S.P. Chase, Apr. 21, 1863, ORN Ser. 1, Vol. 17 at 417. Instructions of G. Welles, Aug. 18, 1862, ORN Ser. 1, Vol. 1 at 418. Lord Lyons to A. Milne, May 11, 1863, quoted in Lord Newton, Lord Lyons: A Record of British Diplomacy, Vol. 1. at 104. S.L. Bernath, Squall Across the Atlantic at 61-62.

With the fall of Vicksburg in July 1863: F.G. Ruffin to L.B. Northrop, Nov. 3, 1862, ORA Ser. 4, Vol. 2 at 159. Report of L.B. Northrop, no date, ORA Ser. 4, Vol. 2 at 193. A.C. Myers to J. Davis, Mar. 4, 1863. L.B. Northrop to J.A. Seddon, Jun. 4, 1863, ORA Ser. 4, Vol. 2 at 574. Indorsement of A.C. Myers, Jul. 4, 1863, ORA Ser. 4, Vol. 2 at 616. J. Gorgas to C. Huse, May 24, 1863, ORA Ser. 4, Vol. 2 at 567-568.

The frustration of the Federal officers: H. Rolando to G.V. Fox, Oct. 3, 1863, ORN Ser. 1, Vol. 20 579-581. The Sir William Peel, U.S. Supreme Court, 72 U.S. 517 (1866).

Captain Rolando's bad luck: J.B. Magruder to S.R. Mallory, Oct. 17, 1863, ORN Ser. 1, Vol. 20 at 579.

Military channels also conveyed: A. Milne to H.H. Bell, Dec. 23, 1863, ORN Ser. 1, Vol. 21 at 108. D.G. Farragut to A. Milne, Feb. 16, 1864, ibid. at 109-110.

In November 1863 a group: E.g., G. Welles to H. Paulding, Dec. 9, 1863 ORN Ser. 1, Vol. 2 at 513. E.g., G. Welles to T.T. Craven, Dec. 10, 1863, ibid. at 517. R.W. Winks, The Civil War Years at 247-254.

The first Federal vessel on the scene: R.W. Winks, The Civil War Years at 247-254.

Chapter 8: The Campaign: 1861-1863

Early in the war, the Confederates: T.O. Selfridge to G. Welles, Aug. 10, 1861 ORN Ser. 1, Vol. 6 at 72. Letter to G. Welles, Aug. 9, 1861, ibid. at 59-60. F.U. Farquhar to J.E. Wool, Sep. 7, 1861, ORA Ser. 1, Vol. 4 at 591. B.F. Butler to J.E. Wool, Aug. 30, 1861, ibid. at 582. Report of W.F. Martin, Aug. 31, 1861, ORN Ser. 1, Vol. 6 at 140. S.H. Stringham to G. Welles, Sep. 2, 1861, ibid. at 121.

Although Federal strategic thinking: C.H. Poor to G. Welles, May 29, 1861, ORN, Ser. 1, Vol. 4 at 187. F. Winslow to J. Pope, Sep. 20, 1861, ORN Ser. 1, Vol. 16 at 683. J. Pope to W.W. McKean, Oct. 3, 1861, ORN Ser. 1, Vol. 16 at 696-697. J. Pope to W.W. McKean, Oct. 13, 1861, ORN Ser. 1, 1 Vol. 6 at 703-705.

The occupation of the Head of Passes: T.T Craven to G. Welles, Feb. 19, 1862, ORN, Ser. 1, Vol. 17 at 137. G. Dewhurst to G. Welles, Mar. 5, 1862, ORN Ser. 1, Vol. 17 at 137-138. W.W. McKean to G. Welles, Apr. 3, 1862, ORN Ser. 1, Vol. 17 at 200-201.

Another Federal Army-Navy expedition: Report of T.W. Sherman, Nov. 8, 1861, ORA Ser. 1, Vol. 6 at 5-6. T.F. Drayton to L.D. Walker, Nov. 24, 1861, ORA Ser. 1, Vol. 6 at 7,10-11. S.F Du Pont to G. Welles, Nov. 11, 1861, ORN Ser. 1, Vol. 12 at 262-264.

About this time, the Confederates abandoned: M. Smith to W.W. McKean, Sep. 20, 1861, ORN, Ser. 1, Vol. 16 at 677-678. J.W. Phelps, to B.F. Butler, Dec. 5, 1861, ORA, Ser. 1, Vol. 6 at 465-468. J.S. Missroon to S.F. Du Pont, Nov. 25, 1861 ORN Ser. 1, Vol. 12 at 326. R.E Lee to J.P. Benjamin, Nov. 29, ibid. at 327-328.

Both the Federals and the Confederates: M. Smith to W. Mervine, Jul. 9, 1861, ORN, Ser. 1, Vol. 16 at 581.

On August 9, 1861, just a: Conference Report (Gulf of Mexico), Aug. 9, 1861, ORN Ser. 1, Vol. 16 at 628.

On the evening of September 16, 1861: M. Smith to W.W. McKean, Sep. 20, 1861, ORN, Ser. 1, Vol. 16 at 677-678. H.W. Allen to Commander of the Massachusetts, Sep. 17, 1861, ibid. at 678-679. J.W. Phelps, to B.F. Butler Dec. 5, 1861, ORA, Ser. 1, Vol. 6 at 465-468. D.G. Farragut to G. Welles, Aug. 20, 1862, ORN Ser. 1, Vol. 19 at 164.

In March 1862 the Federals occupied: S.F. Du Pont to G. Welles, March 4, 1862, ORN Ser. 1, Vol. 12 at 573-575. T.H. Stevens to S.F. Du Pont, Mar. 13, 1862, ibid. at 599-600. H.G. Wright to A.B. Ely, Apr. 13, 1862, ORA Ser. 1, Vol. 6 at 124-125.

In early 1862 the Federals undertook: G.B. McClellan to A.E. Burnside, Jan.7, 1862, ORA Ser. 1, Vol. 9 at 352. W.R. Trotter, Ironclads and Columbiads at 62 and 70. A.E. Burnside to G.B. McClellan, Feb. 10, 1863, ORA Ser. 1, Vol. 9 at 74. A.E. Burnside to L. Thomas, Feb. 14, 1862, ORA Ser. 1, Vol. 9 at 75-81. A.E. Burnside to L. Thomas, Mar 16, 1862, ORA Ser. 1, Vol. 9 at 198. S.C. Rowan to L.M. Goldsborough, Mar. 20, 1862, ORN Ser. 1, Vol. 7 at 111-112. J.R. Weaver II, A Legacy in Brick and Stone at 139-141. J.G. Parke to L. Richmond, May 9, 1862, ORA Ser. 1, Vol. 9 at 283-284. D.W. Flagler to C.T. Gardner, ORA Ser. 1, Vol. 9 at 285-287. A.E. Burnside to E.M. Stanton, Apr. 29, 1862, ORA Ser. 1, Vol. 9 at 273. P. Branch, Fort Macon at 145-146. A.E. Burnside to E.M. Stanton, Apr. 29, 1862, ORA Ser. 1, Vol. 9 at 274.

The Federals also began operations against: Report of T.W. Sherman, Nov. 27, 1861, ORA Ser. 1, Vol. 6 at 192. Report of Q.A. Gillmore, Oct. 20, 1865, ORA Ser. 1, Vol. 6 at 150-163. D. Hunter to E.M. Stanton, Apr. 13, 1862, ORA Ser. 1, Vol. 6 at 133-134.

The capture of Forts Macon and Pulaski: J.R. Weaver II, A Legacy in Brick and Stone at 7. T. Fleming, Beat the Last Drum.

New Orleans, located on the bank: R.V. Remini, The Battle of New Orleans at 169-175. G. Welles to D.D. Porter, Nov. 18, 1861, ORN Ser. 1, Vol. 18 at 3. G. Welles to D.D. Porter, Dec. 2, 1861, ibid. at 3-4. G. Welles to D.G. Farragut, Dec. 23, 1861, ibid. at 4. G. Welles to D.G. Farragut, Jan. 20, 1862, ibid. at 8.

Forts Jackson and St. Philip: Memorandum of J.G. Barnard, Jan. 28, 1862, ORN Ser. 1, Vol. 18 at 15-24.

The forts commanded about 3.5 miles: Memorandum of J.G. Barnard, Jan. 28, 1862, ORN Ser. 1, Vol. 18 at 15-24.

Flag Officer Farragut's squadron: R.U. Johnson, ed., "Battles and Leaders of the Civil War," Vol. 2 at 74. G. Welles to D.D. Porter, Feb. 10, 1862, ORN Ser. 1, Vol. 18 at 25. C.G. Hearn, The Capture of New Orleans at 129. D.D. Porter, "The Opening of the Lower Mississippi" in R.U. Johnson, ed., "Battles and Leaders of the Civil War," Vol. 2 at 29. General Orders, no date, ORN Ser. 1, Vol. 18 at 48-49.

Upriver, the defenses of New Orleans: Testimony of M. Lovell, Apr. 8, 1863, ORA Ser. 1, Vol. 6 at 560. Finding of Court of Enquiry, Dec. 5, 1863, ORN Ser. 1, Vol. 1 at 319.

The Confederates obstructed the river: Testimony of M. Lovell, Apr. 9, 1863, ORA Ser. 1, Vol. 6 at 562 and 564. H.H. Bell to D.G. Farragut, Mar. 28, 1862, ORN Ser. 1, Vol. 18 at 89.

The Federal vessels began crossing the bar: D.D. Porter, "The Opening of the Lower Mississippi" in R.U. Johnson, ed., "Battles and Leaders of the Civil War," Vol. 2 at 29. D.D Porter to G. Welles, Apr. 30, 1862, ORN Ser. 1, Vol. 18 at 362-363. J. Guest to D.D. Porter, Apr. 28, 1862, ORN Ser. 1, Vol. 18 at 377.

The bombardment began on the morning: D.D Porter to G. Welles, Apr. 30, 1862, ORN Ser. 1, Vol. 18 at 364-367. D.G. Farragut to

G. Welles, May 6, 1862, ibid. at 156. C.G. Hearn, The Capture of New Orleans at 197-200.

The Federal squadron advanced at three: D.G. Farragut to G. Welles, May 6, 1862, ORN Ser. 1, Vol. 18 at 156-157. D.D Porter to G. Welles, Apr. 30, 1862, ibid. at 367. C.G. Hearn, The Capture of New Orleans at 234. "The Opposing Forces in the Operations at New Orleans, La." in R.U. Johnson, ed., "Battles and Leaders of the Civil War," Vol. 2 at 73 and 75. P.H. Silverstone, Civil War Navies at 152, 170 and 172. J.H. Russell to D.G. Farragut, Apr. 29, 1862, ORN Ser. 1, Vol. 18 at 224-225 (USS *Kennebec*). C.H.B. Caldwell, to D.G. Farragut, Apr. 24, 1862, ibid. at 225-226 (USS *Itasca*). E.T. Nichols to D.G. Farragut, Apr. 30, 1862, ibid. at 226-227 (USS *Winona*).

Two field artillery emplacements: M.L. Smith to J.G. Pickett, May 6, 1862, ORN Ser. 1, Vol. 18 at 285. D.G. Farragut to G.V. Fox, Apr. 25, 1862, ibid. at 154-155. C.G. Hearn, The Capture of New Orleans at 250-253.

Having captured New Orleans, Flag Officer Farragut: W.H. Roberts, Now for the Contest at 54.

The Federal government was eager: G. Welles to S.F. Du Pont, Jan. 6, 1863, ORN Ser. 1, Vol. 13 at 503.

Secretary Welles stated his desire: G. Welles to S.F. Du Pont, Jan. 31, 1863, ORN Ser. 1, Vol. 13 at 571. Memorandum, Jan. 26, 1863, ibid. at 571-573.

By March the Navy Department had increased: G. Welles to S.F. Du Pont, Jan. 31, 1863, ORN Ser. 1, Vol. 13 at 736-737.

Admiral Du Pont had his doubts: S.F. Du Pont to G. Welles, Jan. 28, 1863, ORN Ser. 1, Vol. 13 at 543.

Forts and shore batteries defended Charleston: Circular of Instructions of R.S. Ridley, Dec. 26, 1862, ORN Ser. 1, Vol. 14 at 102-105. W.H. Parker, Recollections of a Naval Officer 1841-1865 at 332-333.

The Federals expected to encounter obstructions: M.P. Kochan and J.C. Wideman, Torpedoes at 13-25. Sketch of "Devil", Apr. 7, 1863, ORN Ser. 1, Vol. 14 at 93. Testimony of C.P.R. Rodgers in Report of the Secretary of the Navy in Relation to Armored Vessel at 157. J. Rodgers to S.F. Du Pont, Apr. 20, 1863, ORN Ser. 1, Vol. 14 at 43-45.

The nine Federal ironclads weighed: S.F. Du Pont to G. Welles, Apr. 15, 1863, ORN Ser. 1, Vol. 14 at 6-7. Compare J. Johnson, The Defense of Charleston Harbor at 32 (torpedoes not deployed before July 1863) with Examination of M.M. Gray, Apr. 11, 1865, ORN Ser. 1, Vol. 16 at 412 (torpedoes were deployed starting in February 1863). S.F. Du Pont to G. Welles, Apr. 15, 1863, ORN Ser. 1, Vol. 14 at 6-7.

The intense firing lasted about 40 minutes: S.F. Du Pont to G. Welles, Apr. 15, 1863, ORN Ser. 1, Vol. 14 at 7. S.R. Wise, Gate of Hell at 30. W.H. Echols to D.B. Harris and enclosures, Apr. 9, 1863, ibid. at 87, 91-92 and 94-95. S.R. Wise, Gate of Hell at 30.

Although Admiral Du Pont said: S.R. Wise, Gate of Hell at 30-31. S.F. Du Pont to G. Welles, Apr. 15, 1863, ORN Ser. 1, Vol. 14 at 6. A.C. Rhind to S.F. Du Pont, Apr. 8, 1863, ibid. at 23. J. Downes to S.F. Du Pont, Apr. 13, 1863, ibid. at 21-22. P. Drayton to S.F. Du Pont, Apr. 8, 1863, ibid. at 10. J. Rodgers, to S.F. Du Pont, Apr. 8, 1863, ibid. at 12. A.C. Stimers to G. Welles, Apr. 14, 1863, ibid. at 42. D. Ammen to S.F. Du Pont, Apr. 14, 1863, ibid. at 15. L.A. Beardslee to D.M. Fairfax, Apr. 8, 1863, ibid. at 18-19. K.J. Weddle, Lincoln's Tragic Admiral at 195.

Admiral Du Pont became more vocal: S.F. Du Pont to G. Welles, Apr. 15, 1863, ORN Ser. 1, Vol. 14 at 7. A.C. Stimers to G. Welles, Apr. 14, 1863, ibid. at 42. G. Welles to S.F. Du Pont, Jun. 3, 1863, ORN, Ser. 1, Vol. 14 at 230.

NOTES

The increased number of blockade runners: D.G. Farragut to G. Welles, Sep. 8, 1862, ORN Ser. 1, Vol. 1 at 431. R.B. Hitchcock to D.G. Farragut, Jan.16, 1863, ibid. at 27-28.

During the period May through August 1863: Stations of Vessels, Mar 1, 1863, ORN Ser. 1, Vol. 20 at 641.

Morris Island was the long strip of land: S.R. Wise, Gate of Hell at 61.

Federal troops under General Quincy A. Gillmore: S.R. Wise, Gate of Hell at 76-79 and 92-118. D.J. Eicher, The Longest Night at 566-568. Q.A. Gillmore to G.W. Cullum, no date, ORA Ser. 1, Vol. 28, Pt. 1 at 12 and 15-21. J.A. Dahlgren to G. Welles, Jul. 12, 1863, ORN Ser. 1, Vol. 14 at 319-321. R. Reed, Combined Operations in the Civil War at 305-307. J.A. Dahlgren to B.W. Wade, June 20, 1864, in Report of the Joint Committee on the Conduct of the War (Miscellaneous) at 3. R. Reed, Combined Operation in the Civil War at 307. E.W. Serrell to Q.A. Gillmore, Sep. 10, 1863, ORA Ser. 1, Vol. 28, Pt. 2 at 235. T.B. Brooks to Q.A. Gillmore, Sep. 27, 1863, ORA Ser. 1, Vol. 28, Pt. 2 at 267. R.S. Ripley to T. Jordan, ORA Ser. 1, Vol. 28, Pt. 2 at 386. W. Tennent, Jr. to D.B. Harris, Aug 13, 1863, ORA Ser. 1, Vol. 28, Pt. 2 at 510. Report of A. Rhett, ORA Ser. 1, Vol. 28, Pt. 2 at 578. P.G.T. Beauregard to S. Cooper, no date, ORA Ser. 1, Vol. 28, Pt. 1 at 82.

On August 17 the Federals commenced: Q.A. Gillmore to G.W. Cullum, no date, ORA Ser. 1, Vol. 28, Pt. 1 at 21-30. J.A. Dahlgren to B.W. Wade, June 20, 1864, in Report of the Joint Committee on the Conduct of the War (Miscellaneous) at 3-4. Addenda No. 3 to P.G.T. Beauregard to S. Cooper, Sep. 24, 1863, ORA Ser. 1, Vol. 28, Pt. 1 at 101 (notes of discussion on Sep. 4, 1863).

The near wall of Fort Sumter had been reduced: J.A. Dahlgren to G. Welles, Sep. 9, 1863, ORN Ser. 1, Vol. 14 at 610. J.A. Dahlgren to G. Welles, Sep. 11, 1863, ibid. at 610-611. J.A. Dahlgren to G. Welles, Sep. 23, 1863, ibid. at 659-60. J.A. Dahlgren to G. Welles,

Sep. 29, 1863, ibid. at .680-681. J.A. Dahlgren to G. Welles, Oct. 1, 1863, ORN Ser. 1, Vol. 15 at 3-4. J.A. Dahlgren to G. Welles, Oct. 2, 1863, ibid. at 4. T.J. Griffin to J.A. Dahlgren, Nov. 7, 1863, ibid. at 103.

Chapter 9: The Campaign: 1864-1865

Circumstance gave Mobile a two-year reprieve: G. Welles to D.G. Farragut, Jan. 20, 1862, ORN Ser. 1, Vol. 18 at 8. U.S. Grant to E.M. Stanton, Jul. 22, 1865, ORA Ser. 1, Vol. 34, Pt. 1 at 8-9 and 11. J.M. McPherson, Battle Cry of Freedom at 722. L.H. Johnson, Red River Campaign at 242-276.

By June 1864 the complexion of the war: J.M. McPherson, Battle Cry of Freedom at 743-750. D.G. Farragut to N.P. Banks, Feb. 11, 1864, ORN Ser. 1, Vol. 21 at 31. E.R.S. Canby to W.H. Halleck, ORA, Ser. 1, Vol. 39, Pt. 1 at 403.

Mobile sat at the northern end: C.G. Hearn, Mobile Bay and the Mobile Campaign at 41. J.R. Weaver II, A Legacy in Brick and Stone at 171-174.

Fort Gaines sat opposite Fort Morgan: J.R. Weaver II, A Legacy in Brick and Stone at 175-177.

The Confederates obstructed western portion: G.J. Rains to J.A. Seddon, Aug. 15, 1864, ORN Ser. 1, Vol. 21 at 567. J.B. Marchand to D.G. Farragut, Feb. 18, 1864, ORN Ser. 1, Vol. 21 at 105-106.

Several shallow passes west of Dauphin Island: J.R. Weaver II, A Legacy in Brick and Stone at 171. J.M. Williams to G.G. Garner, Aug. 7, 1864, ORA Ser. 1, Vol. 39, Pt. 1 at 441. D.G. Farragut to G. Welles, Feb. 28, 1864, ORN Ser. 1, Vol. 21 at 96-97. C.C. Simms to C.A.R. Jones, Mar. 5, 1864, ibid. at 881.

The four Confederate warships in Mobile Bay: P.H. Silverstone, Civil War Navies at 156, 165 and 178. E.g., D.G. Farragut to J.A. Palmer, Mar. 6, 1864, ORN Ser. 1, Vo. 21 at 127-128; D.G. Farragut to T.A. Jenkins, Mar. 17, 1864, ibid. at 144. E.g., D.G. Farragut to G. Welles, May 9, 1864, ibid. at 267-268. J.B. Marchand to D.G. Farragut, Feb. 24, 1864, ibid. at 115. Instructions of J.B. Marchand, Feb. 22, 1864, ibid. at 116. D.G. Farragut to W. Smith, May 21, 1864, ibid. at 291.

The Federals took steps to counter: D.G. Farragut to G. Welles, May 25, 1864, ORN Ser. 1, Vol. 21 at 298. D.G. Farragut to H.A. Wise, Jun. 11, 1864, ibid. at 331-332. D.G. Farragut to G.V. Fox, Jun. 14, 1864, ibid. at 335. O.A. Batsheller to J.H. Strong, Aug. 5, 1864, ibid. at 473. D.G. Farragut to G. Welles, Aug. 20, 1864, ibid. at 438. D.G. Farragut to J. Lenthall, Oct. 19, 1864, ibid. at 441. D.G. Farragut to G. Welles, Oct. 23, 1864, ibid. at 696-670 (and illustration).

By the eve of battle, four monitors: D.G. Farragut to W.H. Shock, Jul. 18, 1864, ORN Ser. 1, Vol. 21 at 377-378. D.G. Farragut to J.S. Palmer, Jul. 18, 1864, ibid. at 378-379. D.G. Farragut to G. Welles, Jul 21, 1864, ibid. at 381. Diagram of Battle, Aug. 4, 1864, ibid. at 404. D.G. Farragut to T.H. Stevens, Aug. 4, 1864, ibid. at 404. C.G. Hearn, Mobile Bay and the Mobile Campaign at 74 and 78-79.

The Federal plan of attack had the four monitors: D.G. Farragut to T.H. Stevens, Aug. 4, 1864, ORN Ser. 1, Vol. 21 at 404.

The attack began on August 3, 1864: E.R.S. Canby to H.W. Halleck, Aug. 6, 1864, ORA Ser. 1, Vol. 39, Pt. 1 at 402-403.

The Federals did not succeed in silencing: A.W. Bergeron, Jr., Confederate Mobile at 139. J.C. Kinney, "Farragut in Mobile Bay" in R.U. Johnson, ed., "Battles and Leaders of the Civil War," Vol. 4 at 388 (editor's note) and 399 (editor's note). D.G. Farragut to G. Welles, Aug. 12, 1864, ORN Ser. 1, Vol. 21 at 418. C.G. Hearn, Mobile Bay and the Mobile Campaign at 117.

On the same morning, five: D.G. Farragut to G. Welles, Aug. 5, 1864, ORN Ser. 1, Vol. 21 at 405. E.A. Denicke to F.W. Marston, Aug. 12, 1864, ibid. at 507-512 (transmitting copies of messages sent and received). C.G. Hearn, Mobile Bay and the Mobile Campaign at 92, 111-112 and 117. J. Friend, West Wind, Flood

Tide at 189. D.G. Farragut to G. Welles, Aug. 12, 1864, ORN Ser. 1, Vol. 21 at 418 and 420. J. Alden to D.G. Farragut, Aug. 6, 1864, ibid. at 445-446.

The siege of Fort Gaines continued: M.D. McAlester to R. Delafield, Aug. 20, 1864, ORA Ser. 1, Vol. 39, Pt. 1 at 410. D.H. Maury to J.A. Seddon, Aug. 6, 1864, ibid. at 426.

Admiral Farragut took a gamble: Indorsement of D.H. Maury, Aug. 8, 1864, ORA Ser. 1, Vol. 39, Pt. 1 at 442.

Before the Civil War began, Wilmington: C.E. Fonvielle, Jr. The Wilmington Campaign at 14-15 and 21. J.R. Weaver II, A Legacy in Brick and Stone at 141-142. W.R. Trotter, Ironclads and Columbiads at 274.

The ship channel leading to the New Inlet: C.E. Fonvielle, Jr., The Wilmington Campaign at 43-45. R. Gragg, Confederate Goliath at 18-21 and illustration. C.B. Comstock to R. Delafield, Jan. 23, 1865, ORA Ser. 1, Vol. 46, Pt. 2 at 215-216.

For all of its size and apparent strength: Report of T.L. Casey, Dec. 29, 1864, ORA Ser. I, Vol. 42 Pt. 1 at 990.

In late September 1864 General Grant: Testimony of B.F. Butler, Jan. 17, 1865, in Report of the Joint Committee on the Conduct of the War (Fort Fisher Expedition) at 6. Testimony of U.S. Grant, Feb. 11, 1865, ibid. at 52. U.S. Grant to B.F. Butler, Nov. 30, 1864, ORA Ser. 1, Vol. 42, Pt. 3 at 760. U.S. Grant to D.D. Porter, Nov. 30, 1864, ibid. at 750. D.D. Porter to U.S. Grant, Nov. 30, 1864, ibid. at 750.

The idea of using a ship loaded: Testimony of B.F. Butler, Jan. 17, 1865, in Report of the Joint Committee on the Conduct of the War (Fort Fisher Expedition) at 3-4. J.M. McPherson, Battle Cry of Freedom at 722, 726-728 and 758-760. E.g., R. Delafield to C.A. Dana, Nov. 18, 1864, ORN Ser. 1, 11 at 207-214.

The Federal gunboats began assembling: B. Bragg to J.B. Sale, Dec. 20, 1864, ORN Ser. 1, Vol. 11 at 784. Extract of Diary of W. Lamb, ibid. at 746. Extract of Diary of Midshipman Cary, ibid. at 375. A.C. Rhind to D.D. Porter, Dec. 26, 1864, ibid. at 226-227. Instructions of D.D. Porter, Dec. 17, 1864, ibid. at 221-222. R.E. Lee to J.A. Seddon, Dec. 24, 1864, ibid. at 362.

The barrage of Fort Fisher began: D.D. Porter to G. Welles, Dec. 26, 1864, ORN Ser. 1, Vol. 11 at 253, 255-256 and 259. C.E. Fonvielle, Jr., The Wilmington Campaign at 129-32. R. Gragg, Confederate Goliath at 64-65. W.H.C. Whiting to A. Anderson, Dec. 30, 1864, ORA Ser. 1, Vol. 42, Pt. 1 at 994-995. D.D. Porter to G. Welles, Dec. 28, 1864, ORN Ser. 1, Vol. 11 at 261. W.H.C. Whiting to J.F. Gilmer, Dec. 23, 1864, ibid. at 360-361. W. Lamb to J.H. Hill, Dec. 24, 1864, ORA Ser. 1, Vol. 42, Pt. 1 at 1003. M. Long to W. Lamb, no date, ibid. at 1007. Abstract Log of the USS *Colorado*, ORN Ser. 1, Vol. 11 at 296-97. Abstract Log of the USS *Minnesota*, ibid. at 304. W. Lamb to J.H. Hill, Dec. 24, 1864, ORA Ser. 1, Vol. 42, Pt. 1 at 1003. W. Lamb to J.H. Hill, Dec. 27, 1864, ibid. at 1006. W.G. Temple to D.D. Porter, Jan. 2, 1865, ORN Ser. 1, Vol. 11 at 287.

The transports with the Federal infantry: B.F. Butler to U.S. Grant, Jan. 3, 1865, ORA Ser. 1, Vol. 42, Pt. 1 at 967. D.D. Porter to G. Welles, Dec. 26, 1864, ORN Ser. 1, Vol. 11 at 257-258. M. Long to W. Lamb, no date, ORA Ser. 1, Vol. 42, Pt. 1 at 1007.

The Federal infantry began landing: B.F. Butler to U.S. Grant, Jan. 3, 1865, ORA Ser. 1, Vol. 42, Pt. 1 at 967-968. R. Gragg, Confederate Goliath at 80-88 and 95-98. C.E. Fonvielle, Jr., The Wilmington Campaign at 145-162 and 170-176. M. Long to W. Lamb, no date, ORA Ser. 1, Vol. 42, Pt. 1 at 1007. Answers by W.H.C. Whiting to questions posed by B.F. Butler, no date, ibid. at 979. B.F. Butler to D.D. Porter, Dec. 25, 1864, ORA Ser. 1, Vol. 42, Pt. 3 at 1075-1076.

"The Opposing Forces at Fort Fisher" in R.U. Johnson, ed., "Battles and Leaders of the Civil War," Vol. 4 at 662.

General Grant called the aborted attack: U.S. Grant to A. Lincoln, Dec. 28, 1984, ORA Ser. 1, Vol. 42, Pt. 3 at 1087. Enclosure accompanying C.B. Comstock to J.A. Rawlins, Jan. 17, 1865, ORA Ser. 1, Vol. 42, Pt. 1 at 975-977. Indorsement of U.S. Grant, Feb. 2, 1865, ibid. at 977. D.D. Porter to G. Welles, Dec. 27, 1864, ORN Ser. 1, Vol. 11 at 261-262. D.D. Porter to G. Welles, Dec. 29, 1864, ibid. at 263-265. W.T. Sherman to D.D. Porter, Dec. 31, 1864, ibid. at 397.

Notwithstanding the failure: G. Welles to U.S. Grant, Dec. 29, 1864, ORA Ser. 1, Vol. 42, Pt. 3 at 1091. U.S. Grant to A.H. Terry, Jan. 3, 1865, ORA Ser. 1, Vol. 46, Pt. 2 at 25. U.S. Grant to B.F. Butler, Dec. 6, 1864, ORA Ser. 1, Vol. 42, Pt. 1 at 971-972. Testimony of G. Weitzel, Feb. 7, 1865, in Report of the Joint Committee on the Conduct of the War (Fort Fisher Expedition) at 80. U.S. Grant to A.H. Terry, Jan. 3, 1865, ORA Ser. 1, Vol. 46, Pt. 2 at 25.

At four in the morning on January 13: A.H. Terry to J.A. Rawlins, Jan. 25, 1865, ORA Ser. 1, Vol. 46, Pt. 1 at 396. Plan of Second Attack on Fort Fisher, ORN Ser. 1, Vol. 11 after 424. General Orders, No. 78 of D.D. Porter, Jan. 2, 1865, ibid. at 426. Special Orders, No. 8 of D.D. Porter, Jan. 3, 1865, ibid. at 427.

As the firing began, nearly 200 boats: A.H. Terry to J.A. Rawlins, Jan. 25, 1865, ORA Ser. 1, Vol. 46, Pt. 1 at 396-399.

At darkness some Federal gunboats: D.D. Porter to G. Welles, Jan. 17, 1865, ORN, Ser. 1, Vol. 11 at 438. R. Chandler to D.D. Porter, Feb. 4, 1865, ibid. at 484. W.H.C. Whiting to R.E. Lee, Feb. 19, 1865, ibid. at 593.

The Federals scheduled the assault: D.D. Porter to G. Welles, Jan. 17, 1865, ORN, Ser. 1, Vol. 11 at 439. R. Gragg, Confederate Goliath at 142-43.

NOTES

The Federals had prepared bags of gunpowder: A.H. Terry to J.A. Rawlins, Jan. 25, 1865, ORA Ser. 1, Vol. 46, Pt. 1 at 398.

At 3:25 Admiral Porter directed: A.H. Terry to J.A. Rawlins, Jan. 25, 1865, ORA Ser. 1, Vol. 46, Pt. 1 at 398-399. D.D. Porter to G. Welles, Jan. 17, 1865, ORN, Ser. 1, Vol. 11 at 439-441. C.B. Comstock to R. Delafield, Jan. 23, 1865, ORA Ser. 1, Vol. 46, Pt. 2 at 215-216. "The Opposing Forces at Fort Fisher, N.C." in "R.U. Johnson, ed., Battles and Leaders of the Civil War," Vol. 4 at 661-662.

Although the Confederate Armies east: S.R. Wise, Lifeline of the Confederacy at 273, 275 and 283.

The analysis of the effects of the blockade: P. Leigh, Trading with the Enemy.

Without the blockade, the Confederacy: D.G. Surdam, Northern Naval Superiority and the Economics of the American Civil War at 33. J.R. Soley, The Blockade and the Cruisers at 153. S.R. Wise, Lifeline of the Confederacy at 25 and 27-28. E.N. Evans, Judah P. Benjamin at 116. J.M. McPherson, "American Victory, American Defeat" in G.S. Borritt, ed., Why the Confederacy Lost at 23-24. E. Lonn, Desertion in the Civil War. J. Gorgas, "Notes on the Ordnance Department of the Confederate Government", Southern Historical Society Papers, Vol. 12 (Jan.-Feb. 1884) 67-94.

General Edward Alexander Porter: E.A. Porter, Military Memoirs of a Confederate at 53-54.

www.ingramcontent.com/pod-product-compliance
Lightning Source LLC
Chambersburg PA
CBHW071230070526
44583CB00017B/2115